Vegetable Garden Soil Science Made Easy

Create a Soil Base for Abundant Harvests in Your Raised Bed, Container, or No-Dig Garden

By: James R. Bright

Copyright © 2022 by James R. Bright

All rights reserved. No part of this publication may be reproduced, distributed, or transmitted in any form or by any means, including photocopying, recording, or other electronic or mechanical methods, without the prior written permission of the publisher, except in the case of brief quotations embodied in critical reviews and certain other noncommercial uses permitted by copyright law. For permission requests, write to the publisher at the address below:

Heath Publishing Company
1525 Carlos Dr.
Greenville, NC 27834

Table of Contents

Introduction 1

Soil Overview 2

Soil Attributes 7

Soil Fertility 9

The Decline 12

How to Preserve Soil's Fertility 15

Soil Formation 19

Soil Profile 23

Soil Orders 26

Soil Composition 32

Soil Texture 35

Soil Structure 41

Soil Color 45

Soil Permeability 46

Soil Porosity 48

Soil Aeration 49

Soil Microbes 57

Soil pH 66

Soil Salinity 71

Macronutrients and Micronutrients 72

Nutritional Monitoring of Crops 76

No Dig Gardening 78
How to Start a No Dig Garden 80
Raised Garden Beds 83
How to Build a Raised Bed with Wood 90
Filling a Raised Bed 93
Planting Ideas for Raised Beds 98
Container Gardening 100
How to Grow Vegetables in Containers 103
Planting Ideas for Container Gardens 109
Home Compost 111
Purchased Compost 117
A Note on Fertilizers 119
Getting Your Soil Ready for Winter 121
Cover Crops 124
Chop and Drop 128
Conclusion 133

Introduction

Soil is an indispensable, non-renewable natural resource that is vital to life. There are links between soil and water flow, water quality, land usage, and plant productivity. Along with giving instructions on what to do in your vegetable garden, this book starts by discussing essential soil concepts, including formation, classification, physical, chemical, and biological properties, fertility, and preservation. Having a broad understanding of soil concepts and these interdependent linkages will help you have a successful vegetable garden.

Basic knowledge of soil science is an essential step that most beginning gardeners neglect to understand. They spend a lot of time learning about their plants but forget to learn about their soil. How you treat your soil can be the difference between bountiful harvests and a vegetable garden that is a weed filled failure. Most new gardeners who quit never understand why their garden failed, and the most common reason is poor soil.

In order to know how to treat your soil, you need to first understand your soil. This book will help you understand your soil's unique ecosystem and apply the concepts you learn to your vegetable garden. You will also learn how to start a no dig, raised bed, or container garden from scratch. Let's get started.

Soil Overview

It takes hundreds of years for the soil to achieve the minimum thickness for most crops. Initially, fluctuations in temperature and water begin to fracture the rocks. The sun's heat cracks them, and water and ice enter the crevices. The rocks are gradually pulverized and dragged away by the rain and wind. This sediment is deposited in the lower regions when the surface is sloped.

Next, tiny plants and mosses grow by inserting their roots into the crevices. When they die and decompose, they add relatively acidic organic material to the soil, which erodes the stones. Small creatures (worms, insects, fungi, and bacteria) increase and multiply, breaking down and transforming dead plants and animals to add minerals that enrich the soil. Thus enhanced, the soil has improved structure and increased porosity. It permits the growth of larger plants, which provide shade, shelter, and food for an even wider variety of plants and animals.

The significance of soil in sustaining life on earth cannot be overstated. Soils provide a substrate for attenuating pollutants, surplus water, groundwater recharge, nutrient cycling, and microbe and biota habitats. Almost all of the food consumed by humans, except that obtained from aquatic settings, is grown on the planet's soils.

Soils also allow humans to create fiber for paper and textiles, to construct building foundations, and to have fuel wood for cooking and heating. Soils are used as ingredients in candies, pesticides, inks, paintings, cosmetics, and medications. Clay soils are used for anything from drilling muds, ceramics, and artwork, to creating glossy finishes on various surfaces of paper goods.

Soil is a vital component of the majority of ecosystems. However, it is frequently taken for granted. Soil productivity influences what an ecosystem will look like in terms of the plant and animal species it can support, making soil the ecosystem's basis. In forest ecosystems, for instance, soils can influence species composition, wood yield, wildlife habitat, abundance, and diversity. In addition to its vital role in supporting water quality and long-term site production, soil's function in forests is also essential.

Since soil nutrients and physical qualities can directly affect crop yields in cultivated fields, soil quality plays a crucial role in crop productivity. In metropolitan regions, the role of soil in lowering runoff through infiltration and nutrient attenuation is vital. It is easy to overlook the usefulness of soil until the soil's quality deteriorates and the essential functions it formerly offered are lost.

The Natural Resource Conservation Service (NRCS) defines soil as "a natural body composed of solids (minerals and organic matter), liquids, and gases that occurs on the land surface, occupies space, and is characterized by one or both of the following: horizons, or layers, that are distinguishable from the initial material as a result of additions, losses, transfers, and transformations of energy and matter, or the ability to support rooted plants in a natural environment."

Soil is defined by the Soil Science Society of America (SSSA) as "(i) The unconsolidated mineral or organic material on the immediate surface of the Earth that serves as a natural medium for the growth of land plants. (ii) The unconsolidated mineral or organic materials on the surface of the Earth that has been subjected to and shows effects of genetic and environmental factors of: climate (including water and temperature effects), and macro- and microorganisms, conditioned by relief, acting on parent material over a period of time."

Minerals, organic particles, microscopic plant and animal species, air, and water make up soil. Soil is a thin layer that evolved extremely slowly over decades due to the dissolution of the surface rocks by water, temperature fluctuations, and wind. Plants and animals that live and die in and on the soil are decomposed by microbes, converted to organic matter, and incorporated into the ground.

The minerals in the soil are derived from the decomposing bedrock. Additionally, wind and water can transport them from other degraded regions. Organic matter in the soil results from the breakdown of dead plants and animals. It has a significant capacity for water storage and is rich in minerals.

There are two types of microorganisms in soil: those that digest organic materials (insects and worms) and those that do so by releasing nutrients (fungi and bacteria). They reside in the soil and, in addition to aiding in the recycling of organic materials by plants, they contribute to the pulverization of rocks. Earthworms and insects create pores, or holes between soil particles, which promote aeration, water retention, and root development. When it comes to gardening, the distribution and size of pores are crucial. A soil with excessive tiny pores is dense, heavy, and wet, with poor root development. Soils with too many big pores are loose and dry out rapidly. Small pores make it more difficult for a plant to absorb water.

Soil microbes and plants cannot survive without water. Plants utilize it for tissue maintenance, nutrient transport, respiration, and nourishment. The roots take soil water and use it in the photosynthesis process. The breakdown of minerals and organic substances in water makes it easier for plants to absorb them. When soil water is insufficient, plants cease growing and ultimately perish. A surplus of water displaces the soil's air. In addition to water, air is also vital because it allows the roots to breathe oxygen. It is also the nitrogen source that bacteria convert into a form that plants can use.

Thousands of life forms increase on the ground, most invisible to human senses. One acre of fertile land may hold over 121 million small invertebrates, including insects, spiders, earthworms, and other small animals. A teaspoon of soil may contain one million bacteria, hundreds of thousands of yeast cells, and tiny fungi.

All of the soil's constituents are significant in and of themselves, but what is most essential is the proper balance between them. Microorganisms and organic materials provide and release nutrients and link mineral particles together. This allows plants to breathe, absorb water and nutrients, and establish roots. Additionally, earthworms, bacteria, and fungi make humus, which is dark, organic material that forms in soil when plant and animal matter decays. . Humus is a stable form of organic matter which holds water and nutrients and aids in erosion prevention.

Sustainable soil management should promote the activity of microorganisms while maintaining or supplying suitable amounts of organic matter.

Soil Attributes

The features of each soil are contingent on several variables. The essential factors include the type of rock that formed them, their age, the topography, the temperature, the vegetation, the animals that live there, and human-induced changes.

The size of the soil's mineral particles impacts its physical features, including texture, structure, water drainage capacity, and aeration. The granules in sandy soils are more prominent. These are loose and easy to manipulate, but the furrows disintegrate, and water infiltrates rapidly. They have limited plant-usable nutrient stores.

The granules in silt soils are of intermediate size, are dense, and they contain little nutrients. Clay soils are composed of minute particles. They are dense, do not drain or dry out readily, and contain abundant nutrients. They harden and form lumps when they dry. When extremely dry, they are fruitful yet challenging to cultivate.

The components of loam soils are sand, silt, and clay. They are prolific, and when they dry, they break apart into little clumps. A loamy soil with a balanced mineral composition is an agricultural soil with adequate nutrient reserves and is easy to work with. Despite being permeable, it retains moisture. Loamy soils are ideal for most garden plants because they hold plenty of moisture but also drain well so that sufficient air can reach the roots.

Water retention and puddle formation are enhanced when the spaces between soil particles are exceedingly small. The presence of organic matter helps water to permeate and infiltrate slowly, allowing the roots to take advantage of it more effectively. In turn, organic matter restricts nutrient loss and facilitates plant uptake.

The structure of soils is not consistent; they are composed of layers that vary in particle size and composition. The surface layer is more compact, quickly dries, and is inhabited by a small number of species, mainly earthworms. Below it, microbes and nutrients accumulate in the humus.

The soil's chemical characteristics are determined by the proportion of its various mineral and organic constituents. Nitrogen, phosphorus, potassium, calcium, and magnesium must be abundant and in proper balance. In addition to other elements, carbon, oxygen, and hydrogen are always present in organic matter. By decomposing and digesting dead plants and animals, microbes liberate nutrients that can be reused.

Together with climatic variables, the physical and chemical qualities of the soil dictate which plants and animals can grow and how the land should be farmed.

Soil Fertility

We are aware that plants require water and specific nutrients to thrive. They absorb them through their roots from the soil. Soil is fertile if it has the needed nutrients or chemical elements for plant growth.

Carbon, hydrogen, and oxygen, among others, are obtained by plants from the air and water. Other necessary elements are present in the soil. Those consumed in the largest quantities are considered essential macronutrients since plants require them in substantial proportions. The six macronutrient elements are nitrogen, phosphorus, potassium, calcium, magnesium, and sulfur. The elements originate from the rocks that formed the soil and the organic materials broken down by microbes. Nutrients must always be present in the appropriate quantities and ratios.

A soil is fruitful if and only if:
- Its uniformity and depth allow for healthy root development and establishment.
- It supplies the necessary nutrients for plant growth.
- It is capable of absorbing and retaining water, making it accessible to plants.
- It has adequate ventilation.
- It does not contain harmful chemicals.

Soils that are naturally vegetated keep their fertility. The forest is an example: the roots of the trees sustain the earth, while the foliage of the treetops cushions the effects of rain and wind. Together with dead animals and their droppings, dried leaves fall, decay and are digested by microbes to generate humus. Humus is a natural organic fertilizer that improves soil quality, increases surface porosity, and absorbs and retains water slowly. Therefore, the soil retains its moisture for longer, water does not flow off its surface, and there is no land dragging.

The shade of the trees permits the growth of ferns, orchids, mosses, and lichens, which cannot thrive in direct sunlight. Numerous insects and birds feast on its fruits and contribute to plant multiplication by pollinating blooms and dispersing seeds. Meadows of low, bushy grasses help protect the soil. Showers and winds reach the ground through the leaves, which reduce their impact, and the tangled roots support the earth. The steady flow of organic materials has resulted in humus-rich soil.

Agricultural land gradually expends its nutrients and is more susceptible to soil loss. When soil is plowed, its resistance to wind and water erosion is diminished. The decline is exacerbated on sloping ground not covered by windbreaks and hedges composed of trees and bushes.

In addition, the harvested product is generally consumed as food or as a raw resource rather than returning to the soil to replenish it. If no action is taken to recover lost fertility, the ground gets depleted after several years of continuous agriculture.

To continue our productivity, the foundation of our progress, we must preserve soil. Numerous factors contribute to its deterioration, but erosion, pollution, overexploitation of pastures, and deforestation are the most significant.

The Decline

Soil is a thin layer that lies on a bedrock base. This layer took centuries to build but can be destroyed within a few years if care is not taken with its use. Originating from bedrock, soils grow one inch over a period of one hundred to several hundred years, depending on climate, vegetation and other factors. However, land can rapidly deteriorate and become unproductive.

Additionally, just 12% of the earth's surface is easily arable, and just 38% of the earth's surface is potentially arable (World Bank). More significant numbers of difficult-to-work-in places exist. There are several potential obstacles: drought owing to a lack of precipitation, frigid temperatures, non-fertile soils due to a lack of mineral nutrients or because they contain too much salt, terrain that is perpetually covered in snow or ice, and land with a very steep slope.

Loss of fertility, pollution and the disappearance of the soil itself due to erosion are all threats to the earth. Frequently, the loss of fertility or contamination causes the destruction of plants and the quick deterioration of exposed soil.

We know that for plants to grow, they require nutrients to consume. Plants absorb vital elements for their development through their roots, including nitrogen, phosphorus, potassium, magnesium, calcium, and sulfur. These minerals are diminished by agriculture. If they are not replenished with compost and organic matter, the soil's fertility will drop until it is depleted.

We also know that some things are poisonous to plants. Pollution is a type of soil deterioration caused by chemical pollutants hazardous to the health of plants, animals, and humans. It can be caused by irrigation water that has been contaminated by leaking sewage or septic systems, as well as by mining or industrial waste. In addition, pollution can be caused by some insecticides and herbicides, which results in the destruction of non-harmful species and can even be harmful to human health.

Soil erosion is also known as dragging and soil particle loss. It is the result of water and wind acting on vulnerable areas. Raindrops strike the ground with force, gradually destroying its structure. As the water drains, it removes soil particles and nutrients and carries them to lower locations. Streams and rivers erode the banks of the surrounding land. The towed debris settles and fills riverbeds and reservoirs, which increases the likelihood of floods.

The wind also carries particles of fertile soil, particularly when recently tilled or during periods of drought, causing dust storms in some locations. The soil is maintained by the vegetative layer that covers it. The leaves mitigate the effects of precipitation, solar heat, and high winds on the ground, while the roots provide stability. Fallen leaves provide a protective layer and aid in the development of humus.

Limiting the amount of vegetation diminishes the contribution of organic matter and the density of the soil-holding roots. The activity of microorganisms and the fertility of the soil decline. Similarly, it loses porosity and structure, becoming more susceptible to erosion. When the soil becomes weaker and the vegetation that grows on it and helps stabilize it is diminished, rain and wind erosion rises. Destruction of forests, improper tilling, and excessive animal trampling on limited soil (overgrazing) further exacerbate erosion.

Topsoil erosion occurs at a faster rate on vulnerable slopes. The water cannot penetrate the earth's surface and instead flows over it while carrying earth particles. The liquid then concentrates in jets that dig a few millimeters deep furrows through which it flows more quickly. At this point, erosion has already caused significant damage. Still, it can be controlled by constructing stone barriers, cultivating terraces perpendicular to the slope, planting ground-covering grasses, and constructing drainage ditches.

If proper steps are not taken, the furrows will converge, their flow will grow, and they will carve out collapse-prone ravines. As erosion advances, gullies form on the soil, and the fertile layer disappears. In locations with low elevation, flooding increases. The delivered dirt is cleaned to remove nutrients and then combined with small stones. This sediment creates a sterile layer over the valley's ground, which is detrimental to agriculture.

How to Preserve Soil's Fertility

The amount of cultivable land we have is limited and must be utilized with care and conservation methods. Proper soil management aids in its preservation, restoration, and enhancement. To ensure good harvests for many years, it is essential to understand what erosion is and how it takes place. We must be familiar with and employ cultivation techniques that avoid soil erosion and maintain its fertility.

The techniques used to prevent erosion hold the soil together, decreasing the influence of water and wind so that it is not carried away. We can combat the decline in fertility by replacing the ground with the nutrients and organic matter that crops and erosion remove. Not only does the soil's organic matter feed it with nutrients, but it also makes it more porous, allowing it to retain water and be more aerated.

Regeneration of organic matter can occur naturally if the soil is allowed to rest, and new plants are anticipated to grow. However, it is also feasible to replenish the earth with compost, livestock manure, or crop leftovers. Another option is using green manures, (vetch, alfalfa, clover, ryegrass, buckwheat, oats, etc.), which are crops that are not harvested since they are used to replenish the soil. They are buried during the blossoming season when they have accumulated the most nutrients.

Legumes can also be planted to increase nitrogen in the soil. Certain plants classified as legumes, including beans, chickpeas, broad beans, alfalfa, clover, soybeans, and acacias, have nodules on their roots that contain bacteria that absorb nitrogen from the air and fix it in the soil. In this approach, other species utilize nitrogen as nutrition.

Chemical fertilizers can be used but should always be used with moderation and prudence. Before growing plants, it is vital to determine what minerals are lacking in the soil and provide them in the appropriate quantities. Fertilizers can damage crops and kill beneficial soil microorganisms if applied in excess. We must remember that fertilizers are chemical compounds that contain plant nutrients but do not increase soil quality since they lack organic matter.

Crop rotation and intercropping are practices that help preserve fertility by refilling the soil with nutrients and organic matter lost via crop production and erosion. Crop rotation is the annual planting of different crops on the same plot of land. Each species utilizes other nutrients with greater vigor, and its roots grow to varying depths. Thus, as a crop utilizes particular nutrients, the nutrients used by the preceding crop are replenished. This cycle also aids in reducing pests, as the following year, they cannot locate the veggies they specifically target.

Intercropping is creating associations of crops and the concurrent planting of different plant types. For example, if you sow corn, beans, and squash:

- Each crop takes the nutrients it needs independently of the others.
- The corn serves as a foundation for the beans to climb.
- The bean, a legume, enriches the soil by fixing nitrogen.
- The squash shades the earth, conserves moisture and discourages weed growth.

In addition to enriching the soil, vegetative cover protects it from erosion, particularly during the rainy season. It also keeps the ground from drying out during the dry season by limiting water loss through evaporation. It is possible to cultivate cover crops between agricultural cycles. Additionally, using stubble (the stalks of plants left sticking up out of the ground after the crop is harvested) as a cover aids in erosion prevention, weed control, and enhances organic matter and fertility. A key to erosion control is to never allow bare soil. When the soil is not growing something, it should be covered by organic material.

Another key to erosion control is minimal disturbance. Minimum tillage restricts soil cultivation to the furrows where seeds will be planted. The remaining acreage is undisturbed. This form of tillage preserves the soil's structure, lessening the drag generated by rain and wind.

In agricultural settings with livestock, it is also essential to prevent overgrazing. When cattle concentrate, the soil is compacted by their frequent trampling. By selectively consuming their preferred grasses, these grasses gradually disappear. However, livestock can also be used to add fertility back to soil through their manure, as long as their pasture is rotated properly and is not overgrazed.

Because water rushes with greater force in sloping regions, soil erosion is more severe. To prevent water and wind from transporting soil particles, we can employ procedures that, despite their simplicity, are very effective. When cultivating sloped soils, cultivation must be conducted perpendicular to the slope or along contour lines. Thus, each groove functions as a barrier that slows water flow. By decreasing surface drainage, the topsoil is preserved. In addition, by reducing the velocity of the water, we will be able to utilize it more efficiently.

To protect the soil from wind and rain-caused erosion in sloping areas, barrier construction is required. Live barriers can be produced by edgings of evergreen trees and shrubs with dense growth that run perpendicular to the direction of the wind and the slope of the ground. It is also beneficial to construct stone barriers to prevent water from eroding quickly and dragging soil particles. The residual soil accumulates and is ideal for crop amendments. Diversion channels and ditches permit the collection of runoff water, which can be accumulated there in place (infiltration grooves) or transported off the land to storage tanks.

On lands with a particularly steep slope, the construction of terraces or platform decks can aid in absorbing water, preventing it from pulling the soil and causing erosion. Additionally, flat and deeper surfaces are created, allowing for the cultivation of a variety of crops.

Soil Formation

Through the process of rock weathering, the soil is created. The process by which rocks are broken down into smaller particles when exposed to water, air, or living beings is known as weathering. The process of weathering might be physical, biological, or chemical.

Physical weathering is the breakdown of rocks into smaller particles without a change in their molecular structure. Air and water are physical weathering agents. Wind blown on rocks, intense rainfall, and ocean waves can facilitate the progressive fragmentation of rock particles into sediments, which ultimately produce soil.

Chemical weathering is when chemical reactions within rocks alter their mineral makeup. Hydrolysis, carbonation, oxidation, and hydration are chemical reactions that contribute to weathering. Common types of chemical weathering include the following:

- Hydrolysis: hydrolysis happens when rainwater percolates through rocks, and the hydrogen ion (H+) in water combines with metallic ions in stones, leading to the dissolving of rock minerals.

- Carbonation: Carbon dioxide from the atmosphere and living organisms dissolves in water to form carbonic acid. This acidifies a rock's moisture, resulting in different chemical reactions with rock minerals.

- Oxidation: During oxidation, oxygen in the air combines with iron in rocks to produce iron oxides. This chemical reaction imparts a rusty brown hue to stones.

With biological weathering, living organisms aid in the fragmentation of rocks. The growth or penetration of tree roots and mosses through rocks creates pores that gradually force rocks apart. Animals tunnel through rocks and cause their decomposition. Microorganisms, such as lichen (a symbiotic connection between fungi and algae) emit compounds that decompose rock minerals.

Climate, geography, living organisms, parent material, and time influence soil development. Climate plays a significant role. Patterns of precipitation and temperature influence the formation of soils in a region. Rainfall provides the water necessary for chemical and physical weathering.

Different climates throughout the world result in a variety of soil types and properties. For instance, the soil is more susceptible to erosion, weathering, and leaching in regions with high or heavy precipitation. In such areas, calcium, magnesium, and potassium minerals are frequently replaced by H^+ through rainfall, resulting in a higher prevalence of acidic soils. Alternatively, areas with low precipitation are less susceptible to leaching. In the absence of leaching, calcium carbonates tend to accumulate in the upper soil layers of these places, resulting in high alkalinity. As plants and microorganisms thrive in warmer temperatures, rock is subjected to quick weathering in these warm places, but the rock is weathered gradually in colder climates.

Topography also has a significant effect on soil development. Topography can be defined as the shape of the earth's surface area. The steepness of a terrain enhances its susceptibility to wind and water erosion, resulting in the movement of rock sediments and the deposition of rock minerals in valleys. In these places, the soils in the valleys are darker, richer in organic content, and more conducive to plant growth, whereas the grounds on the hilltops are eroded, less fertile, and unfavorable to plant growth.

Living organisms are crucial to the production of soil. A region with abundant vegetation is typically rich in humus. Humus is produced as soil microorganisms transform fallen leaves and decaying plant materials into organic matter. The microorganisms consume the organic matter's sugars and carbs, leaving behind lipids and waxes in the soil which create the humus.

The majority of soil parent material is worn rock. Parent material might be sedimentary rocks, igneous rocks, or sediments geologically young which overlie bedrocks. The type of soil depends on the material from which it originated. At the beach, the ground is generated from granite rocks eroded into sand. Quartz, a silicate mineral resistant to weathering, is present in granite. Due to quartz's excellent resistance to deterioration, granite disintegrates into sand particles. Similarly, rocks containing feldspar (a low-resistance silicate mineral) are broken down into small clay particles.

Time is an essential element that is tied to all other soil-forming factors. When climatic, topographic, or biological circumstances are conducive to weathering, soil formation will proceed more rapidly. Time also reveals the duration of other soil-forming components. Typically, regions with young parent material are more fruitful because they have not undergone much weathering and still contain important mineral content. Similarly, areas with older parent material subjected to much more weathering would have less mineral composition. These types of soils derive more of their nutrients from organic materials.

Soil Profile

The soil profile refers to the vertical portion of soil that reveals distinct horizontal strata. Horizon refers to the different or separate layers within the soil profile. The majority of grounds consist of many horizons. Typically, the horizons of a soil profile will mirror the landscape's topography. Measurements of soil color, texture, structure, consistency, root dispersion, effervescence, rock pieces, and reactivity are also used to determine horizon boundaries.

The O horizon, the uppermost layer, is composed mainly of organic material. Typically, forested regions have a distinct O horizon. In other environments, such as a grassland or farmed field, an O horizon may not exist. Erosion and frequent tilling contribute to the absence of organic materials. The O horizon has three major sub-classifications or subordinate distinctions (denoted by a lowercase letter): hemic (Oe), fabric (Oi), and capric (Os or Oa). The hemic layer is made of slightly decomposed decaying matter whose origin is still discernible. The fabric layer is formed of organic material that is somewhat more deteriorated and unrecognizable but not destroyed. The capric layer is composed of wholly decomposed material of unknown origin.

The A horizon is a mineral horizon that forms at or just beneath the surface of the soil. It is generally known as "surface dirt." In addition to the buildup of organic debris, the A horizon may exhibit the presence of a plow pan. A plow pan (or plow layer) is a common feature of soils subjected to conventional plowing in the recent past. Sometimes, the darkness of the A horizon is attributable to the flow of organic stuff from the O horizon. Soils subjected to intensive agriculture will include components that would otherwise comprise the O horizon. These organic components also contribute to the A horizon's organic content, which is higher than other horizons.

The E horizon (eluvial layer) is a typical mineral horizon in forest soils that lacks clay, iron (Fe), and aluminum (Al). Removing the chemicals above and black minerals from the soil particles is known as eluviation. Leaching removes clay, Fe, and Al from the E horizon, resulting in its lighter hue relative to the neighboring strata. Leaching is the passage of water through the soil profile that causes the loss of nutrients from the root zone. The E horizon contains concentrations of minerals that are less prone to leaching, such as quartz, silica, and others.

If present, the B horizon, also known as the "zone of accumulation," lies under the O, A, and E horizons. The B horizon absorbs illuviated elements such as clay particles, Fe and Al oxides, humus (organic matter generated by the decomposition of plant and animal matter), carbonates, gypsum, and silicates leached from the horizons above it. Usually, the B horizon is redder or darker than the neighboring horizons due to the ubiquitous occurrence of Fe and Al oxide coatings.

The A, E, and B horizons together are known as the soil solum. This part of the profile is where most plant roots grow. The soil solum is positioned above the parent material (C and R horizons).

The C horizon, also known as the regolith, is the soil layer often unaffected by pedogenic weathering processes and, as a result, is composed of the partially weathered parent material. The C horizon is the boundary between soil and bedrock. As the upper portion of the C horizon experiences weathering, it may become a component of the superimposed horizons. There is a noticeable difference in soil structure between fully developed B and C layers, which helps to detect the horizon boundary in the field; however, the shift in design may be more subtle in poorly developed soils.

Under the C horizon is the R horizon, which represents the bedrock. Depending on the geographic region, weather circumstances, and landscape position, bedrock may be more than 100 feet deep or as little as centimeters below the earth's surface. Bedrock is a solidified layer of rock material that provides the site's soil qualities. Infrequently, tree roots can create enough stress to fracture bedrock. Biochemical processes are responsible for the majority of deeper bedrock weathering. In contrast to solid rock (i.e., bedrock), the name saprolite/caprock is commonly applied to the layer of newly eroded material.

Not all soil profiles contain identical horizons. Some soils will have O, A, E, B, C, and R horizons, while others may only have C and R horizons. These variations in horizons are what distinguish each soil type. Soil scientists can classify soils using Soil Taxonomy based on the soil's unique properties.

Soil Orders

Numerous techniques for classifying soils according to their morphological and chemical features have been developed globally. Soil Taxonomy, introduced by the United States Department of Agriculture (USDA), is the most popular classification system. This morphogenetic system employs both quantitative criteria and soil genesis themes and assumptions to classify soil.

The Soil Taxonomy System is a hierarchical classification scheme with six levels. The classification levels are as follows, from broadest to narrowest: 1) Order, 2) Suborder, 3) Major Group, 4) Subgroup, 5) Family, and 6) Series. There are currently 12 soil orders, 65 suborders, 344 major groups, over 18,000 subgroups, and over 23,000 soil series.

The defining criteria of the broader classification levels are based on soil-forming processes and parent materials, while the narrower groups consider the arrangement of horizons, colors, textures, etc. Climate has the most crucial role in Soil Taxonomy categorization, followed by parent material and biota. Topography and time are not used to classify soils.

Entisols are the youngest or most recently created soil order. Entisols are characterized by limited profile development, with little or no evidence of horizonation (horizontal layers). Occasionally, Entisols have a weakly formed A or Ap (plow layer) horizon. These soils can be found on steep slopes with significant erosion, in floodplains receiving alluvial deposits, and in various other situations.

Inceptisols are another soil order with fragile profile properties. In 1975, when soil taxonomy was first developed, Inceptisols were referred to as the "wastebasket soil order." These soils generally did not fit within the existing soil orders at the time. As a result of the introduction of additional soil orders, several Inceptisols were reclassified, and the term "wastebasket" is no longer applicable.

Inceptisols exist between no profile development and weak profile development. Due to their limited profile development, Inceptisols are widespread near rivers and streams. Both Entisols and Inceptisols have the potential to grow different horizons and be classed into a different soil order over time. Young soils, such as Entisols and Inceptisols, are more heavily influenced by their parent material than older soils, which are more affected by climate and vegetation conditions.

Gelisols are also "new" soils relative to geologic time, as they formed under cold or frozen conditions. In regions like Canada and Alaska, these soils are frequently connected with permafrost (a layer of earth that is permanently frozen) conditions and cryoturbation (frost churning).

Mollisols are sometimes referred to as prairie soil. Primarily developed beneath grassy prairies, these soils are distinguished by their high organic matter concentration, dark hue, and deep A horizon.

The A horizon must have a minimum depth of 8 inches and a base saturation of at least 50% to be classified as a Mollisol soil. (Base saturation refers to the soil's ability to supply three important plant nutrients: calcium, magnesium, and potassium.) Mollisols are prevalent in the midwestern United States and were initially dominated by native prairies.

Alfisols, which form beneath deciduous forests, are also prevalent in the Midwest of the United States. Alfisols are ordinarily found in humid regions of the world and frequently include an E horizon in their soil profiles. These soils must have a minimum base saturation of 35%.

Spodosols are typically derived from coarse-textured (i.e., sand-rich) and acidic parent materials. Spodosols are generated beneath forest flora, particularly coniferous forests, due to the accumulation of acidic pine needle resins. When pine litter decomposes, highly acidic substances drain through the coarse particles, carrying iron, aluminum, and hummus. Thus, a layer of humus and Fe/Al oxides forms alluvial. Similar to with Alfisols, E horizons are widespread in Spodosol soil profiles. Frequently, a Spodosol will have a white E horizon atop a brilliant red B horizon. Spodosols are found in Wisconsin, Michigan, the northeastern United States, and the coastal plains of the eastern and southeastern United States.

Aridisols are typically found in semi-arid and arid environments. These regions have low yearly mean precipitation. Lack of soil moisture influences soil growth and the weathering process.

Consequently, these soils are predominantly affected by physical weathering, not chemical weathering. Aridisols have a high base saturation percentage of 100 percent. These soils are widespread in the deserts and arid regions of the western United States.

Ultisols can be found in humid and warm places, such as the southeastern United States. The rapid rate of clay mineral weathering and translocation in Ultisols results in the accumulation of clays at the subsurface. They have a base saturation of less than 35% and are less fruitful than Alfisols and Mollisols. However, Ultisols react positively to fertilizer control and are planted worldwide. These soils are often more acidic than Alfisols but less acidic than Spodosols due to their high degree of weathering.

Oxisols are the most severely deteriorated soil order in the United States. Oxisols are so-called because they are oxidized. They are characterized by high clay content and Fe/Al hydrous oxides, which often impart a red color to the soil. Oxisols are found in tropical and subtropical regions, including Hawaii, Puerto Rico, South America, and Africa. Oxisols are typically generated in wetter conditions, but they can also be found in locations that are currently drier than when the soils were formed.

Vertisols are clay-rich soils that lack profile development due to expansion and contraction. These mechanisms mix the ground, preventing the formation of a distinct soil profile. During dry conditions, the soil contracts and fractures up to thirty inches deep and one inch wide. Clays containing smectite, montmorillonite, and vermiculite are present in locations such as southeast Texas and eastern Mississippi in the southern United States, where Vertisols are found.

In 1990, the Andisol order was established. Previously, these soils were classified as Inceptisol. Andisols are volcanic soils of recent origin (young) formed from volcanic elements. They inhabit Hawaii and the northwestern United States, specifically Washington, Idaho, and Oregon.

The only organic soil order in the classification system is the Histosol. Histosols have many sublayers inside the O horizon and include at least 20 percent organic materials. Histosols are found in Wisconsin, Minnesota, the Florida Everglades, and along the Gulf of Mexico, among other locations in the United States and Canada.

Suborders are the next classification following soil orders and classify soil qualities connected with climate meaning. The next classification is the major group, which classifies soils according to the most significant features, such as the kind and order of soil horizons, temperature regimes, and moisture regimes. Then, using subgroups, scientists classify soils further by evaluating the similarity between specific soils and categorizing them accordingly. These grades represent transitions to other orders, suborders, or major groups.

The next classification level is soil families. Members of the same soil family share physical, chemical, and mineralogical features associated with plant growth. A soil series is the lowest level of taxonomy and the most specific soil classification. In addition to physical, chemical, and mineralogical qualities, the soil series categorization also considers management, land-use history, vegetation, topography, and landscape position. Many soil series are named after the place where they were initially discovered.

Soil Composition

The development of soil depends on a lengthy and intricate process of rock degradation, including physical, chemical, and biological processes. As ecological variables, their interaction promotes the dissolution of minerals, which, along with the remains of animals and plants in organic matter, constitute the soil.

In addition to climatic forces, living organisms participate in the mixing of soil chemicals, their horizontal distribution, and the expansion of organic matter to the soil. Animal and plant wastes, as well as their bodies when they die, are the only sources of organic matter in the ground. The organic matter provides essential components, alters the soil in various ways, and enables the formation of diverse fauna and flora that could not exist otherwise.

In addition, the organic matter absorbed into the soil stores more solar energy than the inorganic matter from which it was created. Living organisms contribute to soil development by providing not just materials but also potential and kinetic energy.

Due to intensification and competitive use that characterizes the use of land for agricultural, forestry, pastoral, and urban purposes, as well as to satisfy the growing population's demand for food production, energy extraction, and extraction of raw materials, the natural surface of productive land is limited and under increasing pressure. The productive capacities of soils and their contribution to food security and maintaining essential ecosystem functions must be acknowledged and valued.

Various processes continually modify soils (human action, erosion, climatic changes). Soils consist of four primary components:

- Minerals with varying sizes.
- Organic materials derived from decomposing plants and animals.
- Water that enters the soil through its pores.
- Air that fills the soil's pores.

Minerals or the inorganic components constitute 40 to 50 percent of the overall volume of the soil. The formation of minerals results from the decomposition (erosion or in-situ weathering) of preexisting or parent rocks (igneous rocks, sedimentary rocks, and metamorphic rocks). Particles of gravel, sand, silt, and clay make up the solid portion of the soil, which is composed of minerals and rock fragments.

Living organisms (microorganisms, plants, etc.) and dead organisms (plant residues, dead microorganisms, animal excrement, etc.) comprise the soil's organic component, accounting for approximately five percent of the total volume of the soil. Soil rich in organic matter has a great capacity to hold water and essential nutrients, and is, therefore, conducive to crop production.

The water in the soil is stored in the spaces between the particles that make up the soil (pores). It comprises between 20 and 30 percent of the soil's total volume. Water's significance lies in the soil's high transport capacity, providing essential nutrients for soil life, and to allow biochemical breakdown. The percentage of water held by the soil will rely mainly on the various soil types, with clay soil having the maximum water retention capacity and gravelly-sandy soils having the lowest water retention.

Gases are another of the soil's fundamental components. Gases represent the air that fills the crevices between the soil's solid particles (porosity). Air typically comprises 20 to 30 percent of the soil's total volume. Oxygen is necessary for the respiration of roots and microbes, which contributes to maintaining plant growth.

Soil Texture

The soil texture is the sum of sand, silt, and clay proportions. The soil's surface is a relatively permanent property that determines the biophysical parameters of the soil. Long-term relationships exist between the soil texture, fertility, and quality of the soil. The soil texture is related to soil porosity, which governs the soil's water-holding capacity, gaseous diffusion, and water flow, influencing the soil's health.

The gaseous diffusion and water infiltration stimulate the survival of microbial propagules (a vegetative structure that becomes detached from a plant and can create a new plant). Gaseous diffusion and water infiltration also provide the availability of moisture and oxygen for microbial development. Gaseous diffusion and water infiltration vary with soil texture. They impact the soil carbon dioxide generation in clay loam soil 50% more than they do in sandy soil.

Additionally, soil texture impacts the rooting system and thus controls the soil carbon dioxide outflow. Coarser textured soils with lower water storage capacity, unsaturated hydraulic conductivity (a measure of soil's water-retaining ability when soil pore space is not saturated with water), and lower fertility have slower root growth than finer textured soils. In addition, soil texture impacts the extent of root litter decomposition and rhizosphere (the region of soil in the vicinity of plant roots) microbial respiration, with clay soil having a more significant impact than sandy loam soil.

The percentage of sand, silt, and clay that the soil contains determines its texture. Texture refers to particle sizes smaller than 2 mm and subtle elements. Particles with diameters more than 2 mm are known as gravel. The size of these particles or tiny components determines their limits, which vary based on the granulometric classification employed (the two most frequent classifications are the international and the USDA).

We will concentrate on the USDA categorization, which is the most generally used, and which divides particle sizes as follows:

- Sand fraction: between 2 mm and 0.05 mm. Within sand, it is possible to distinguish between very coarse sand (from 2 mm to 1 mm), coarse sand (from 1 mm to 0.5 mm), medium sand (from 0.5 to 0.25 mm), fine sand (from 0.25 to 0.10 mm) and very fine sand (from 0.10 to 0.05 mm).
- Silt fraction: between 0.05 and 0.002 millimeters
- Clay fraction: particles less than 0.002 mm in diameter

Once the soil's percentages of sand, silt, and clay have been determined, its textural class may be determined. The USDA provides a textural diagram or texture triangle tool for this purpose. Introducing in the figure the data gathered from sand, silt, and clay, we classify the soil's principal texture classes as follows:

- Clay-based soils
- Sandy soils
- Silty soils

- Loamy or equilibrated soils

In addition, several combinations between them are possible, including sandy-clayey, silt-clayey, loam-sandy, loam-silt, etc.

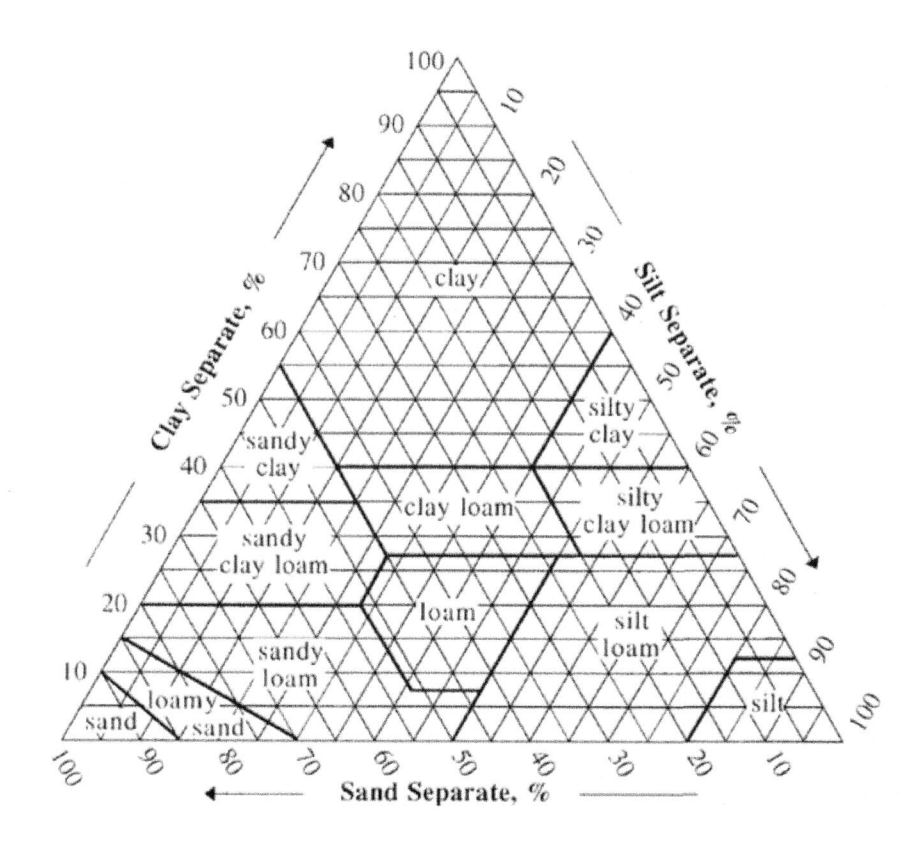

USDA's Textural Triangle Tool

The texture is a variable property throughout the depth of the soil; it is not the same in the upper horizons as it is at greater depths (the same as other soil properties). For this reason, conducting two analyses at different depths may be preferable when establishing crops with deep roots.

The Bouyoucos method is used in the laboratory to determine the percentages of sand, silt, and clay to choose the texture. Using this method, a hydrometer measures the specific gravity of the soil suspension at the center of its bulb. The specific gravity depends upon the mass of solids present, which in turn depends upon the particle size. If you are already getting your soil tested, you should get a texture analysis along with the test.

There are also indirect processes that can provide us with an approximation of the soil's texture, such as forming a small wet ball with the soil we wish to investigate. The features of this small ball allow us to estimate the proportions of the three fractions. If the ball is very flexible, it will include more clay, and if it is rough, unmanageable, and crumbles, it will contain more sand.

Certain characteristics are provided by each textural class to the soil. Physical features like drainage and aeration capacity, as well as chemical characteristics like fertility, are influenced by the soil's texture.

Soils composed of clay particles are highly flexible, heavy, and challenging to work with. Due to their high cation exchange capacity (the soil's ability to supply three important plant nutrients: calcium, magnesium and potassium), they are typically exceedingly fruitful and retain more water than other species (although not all types of clay have a high cation exchange capacity). The most significant negative characteristic of clayey soil is its low water infiltration, as it has limited permeability and is susceptible to frequent and lengthy floods, which can hinder the establishment of crops.

Soils composed of sand particles are light and easy to work with. Due to the significant infiltration of soils with high permeability, flooding is very unlikely to occur. The negative qualities of these soils stem from their low fertility and how quickly they dry out.

Soils composed of silt particles are fertile soils that hold nutrients and water better than sandy soils and are easier to work with than clay soils. However, silty soils have minimal aeration and a propensity to become compact and hard, creating crusts that hinder water infiltration.

Soils having a combination of clay, sand and silt are well-balanced soils with favorable qualities of each texture type. These soils are referred to as loam. Ideally, the ground used to grow crops should have a loamy texture.

Understanding the nature of texture, its various classes, and the features of each is essential for the proper development of crops. Generally, before establishing a crop, a soil analysis must be performed, where one of the most critical parameters is the texture. This is because it is possible that the plant you have chosen to cultivate does not do well with certain textures or that it is necessary to perform some prior work to avoid the influence of the surface, such as making ridges or carrying out drainage works to prevent water accumulation in very clay soils.

It is also good to know the texture of the soil for irrigation purposes. If you know the exact surface, you can more precisely calculate your irrigation schedule. For example, sandy soil would require more frequent irrigation but less water than clayey soil, which would require less regular irrigation, but more water.

Your soil fertility is another critical piece of information provided by texture analysis. Typically, the higher the concentration of clay in the soil, the greater its fertility. The kind of soil texture also impacts whether organic matter will be destroyed quickly (sandy-textured soils) or slowly (clay-textured soils) or somewhere in between (silty-textured soils).

Soil Structure

The structure is how soil particles combine to create aggregates (groups of soil particles). Soils with a spherical structure (rounded aggregates), laminar structure (aggregates in sheets), colorful structure (in the shape of a prism), blocky structure (in blocks), and granular structure (in grains) are identified based on this property.

The organization of sand, silt, and clay particles determine the soil's structure. When particles are aggregated, they take on the appearance of larger particles and are referred to as aggregates.

Soil structure can be divided into grades. The degree of structure expresses the difference between the cohesiveness within the aggregates and the stickiness between them. Because these characteristics change with soil moisture content, the degree of soil structure should be measured when the soil is neither too wet nor dry. There are three primary degrees of soil structure as follows:

1. No Structure: a condition in which there are no visible aggregates or there is no natural ordering of lines of weakness, such as:
    a. Agglomerate (coherent) structure in which the entire soil horizon appears cemented into one large mass;
    b. Single grain structure (no coherence) in which individual soil particles show no tendency to clump together, such as in pure sand;

It is poorly formed by aggregates that are barely perceptible and indistinct. When the materials are withdrawn from the profile, they are shattered, resulting in a mixture of a few intact aggregates, numerous fragmented aggregates, and a great deal of non-aggregate debris.

2. Moderate Structure: Defined by well-formed and identifiable aggregates of moderate duration that are visible but unclear in undisturbed soils. When extracted from the profile, the soil material fragments into a variety of different entire aggregates, some fragmented aggregates, and a small amount of non-aggregate material.

3. Robust Structure – Defined by well-formed and identifiable aggregates that are resilient and visible in undisturbed soils. Extracted from the profile, the soil material predominantly consists of complete aggregates, with some fragmented aggregates and little or no non-aggregate particles.

The structural class describes the average size of the individual aggregates. The following five structural classes can be distinguished:

- Very fine or very thin
- Fine or thin
- Medium
- Thick
- Very thick

The structure type describes the shape or arrangement of the individual aggregates. The four most common structure types are as follows:

1. Granular and crumb structures: comprise sand, silt, and clay particles clustered into roughly spherical grains. Granular and crumb structures offer the most pore space of any structure, facilitating the circulation of water. Typically, they are located in the A horizon of soil profiles.

2. Blocky and subangular blocky structures: These are soil particles organized into almost square or angular blocks with more or less prominent edges. This structure promotes good drainage, aeration and root penetration. Usually, these structures are found in the B horizon.

3. Prismatic and columnar structures: These soil particles have formed vertical columns or pillars divided by tiny but noticeable vertical fractures. Water circulation is hampered, and drainage is inadequate. They are typically found in the B horizon when clay is present.

4. Platy structure: Comprised of soil particles collected in horizontally accumulating sheets or thin layers. Frequently, the sheets overlap, making it extremely difficult for water to circulate. The platy structure has the least amount of pore space and is common in compacted soils. This structure is observed in forest soils, as part of the A horizon, and in clay-layered soils.

There are also structureless soil types, including:

1. Massive structure: Large cohesive masses of clay.

2. Single grain structure: Soils with no true structure, such as a loose sand with little or no attraction between the grains of sand.

Soil Color

The color of soil depends on its constituents and can be used as an indirect indicator of particular features. The hue varies with the relative humidity. Red shows the presence of iron and manganese oxides; yellow, hydrated iron oxides; white and gray the presence of quartz, gypsum, and kaolin; and black and brown the presence of organic debris. Because of the benefits of organic matter, the darker soil is, the more productive it will be for crops.

The color of the soil might provide crucial information about other soil characteristics. For instance, soils that are grayish in hue and include "specks or stains" indicate inadequate aeration. Darker surface horizons absorb more radiation and, as a result, have greater temperatures than lighter soils. The measuring of soil color utilizes a standardized system based on the "Munsell Color Chart." In this table, the following three color components are measured:

- Tone (color) (normally reddish or yellowish in soils)
- Intensity or brilliance (chroma)
- Brightness value

Soil Permeability

Permeability is the soil's ability to convey water and air. The more filtration, the greater the soil's permeability. Numerous factors affect the soil's permeability. Occasionally, there are highly localized characteristics, such as fissures and gullies, which make it challenging to calculate accurate permeability values.

A thorough investigation of soil profiles is a crucial validation of permeability measurements. Based on observations of soil texture, structure, consistency, color and color spots, layering, visible pores, and depth of impermeable layers such as bedrock and clay layers, it is determined whether the permeability measurements are likely to be representative.

The soil is made up of various horizons; generally, each has different physical and chemical properties. To assess the overall permeability of the soil, each horizon must be investigated individually. The permeability of soil is correlated with its texture and structure. Regarding the seepage rate (the passage of water into the soil) and the percolation rate (movement of water through the soil), the size of soil pores are of utmost importance. The size and quantity of pores are intimately related to the soil's texture and structure and impact its permeability.

In general, as indicated below, the finer the texture of the soil, the lower its permeability:

- Sandy loam: 2.5 cm/per hour
- Franco (medium texture): 1.3 cm/per hour
- Clay loam: 0.8 cm/per hour
- Silty clayey: 0.25 cm/per hour
- Clay: 0.05 cm/per hour

Soil Porosity

A soil's porosity, or system of empty spaces and pores, is the result of its texture and structure. There are two distinct types of soil pores: macroscopic and microscopic. Macroscopic pores have larger dimensions and are typically filled with air; hence, water moves through them rapidly, propelled by gravity. In contrast, microscopic pores are mainly filled with water retained by capillary forces.

Sandy soils are rich in macroscopic pores, which permit the rapid passage of water but have a shallow ability to hold water. In contrast, clay soils are rich in microscopic pores and may exhibit poor aeration but have a great capacity to store water. Soil compaction decreases porosity as bulk density increases.

Porosity often varies among the following ranges:

- Light soils: 30 to 45%
- Medium soils: 45 to 55%
- Heavy soils: 50 to 65%
- Peaty soils: 75 to 90%

Soil Aeration

The soil's aeration capacity is one of the most important factors for plant growth, yet it is frequently underestimated. Most people have learned that plants take up carbon dioxide from the air to be used in photosynthesis and produce oxygen as a byproduct of that process, but less well known is that plants also need oxygen.

Plants, like animals, have active metabolisms, fueling all bodily activities. For this, almost all organisms need oxygen, which interacts with glucose (the breakdown of organic compounds) to produce energy, and this complex process produces carbon dioxide (and water molecules) as a byproduct. Most of the carbon dioxide is used by the plant for photosynthesis, but any excess needs to be eliminated.

A soil deficient in oxygen may prevent plant growth and eventually lead to death. Soil aeration gives air to the subsoil by transferring oxygen and carbon dioxide between the soil's pores and the atmosphere. Soil aeration prevents crops from being oxygen-starved and decreases hazardous carbon dioxide levels in the ground air if they begin to rise.

Plant roots require ambient oxygen to breathe, and the glucose-oxygen interaction provides them with the energy they need. In insufficiently aerated soil, oxygen-starved roots perish because they cannot breathe properly. The effects of soil aeration extend beyond crop development. Soil aeration is required for soil-dwelling aerobic bacteria to oxidize adequately.

To ensure proper soil aeration, it is vital to identify the factors that influence it. Then the gardener can understand how to implement aeration techniques and reduce the harmful effects of poor aeration if needed. The following things affect soil aeration:

- Soil moisture
- Soil texture
- Infiltration properties
- Equipment use
- Organic matter treatments
- Grazing use

These variables can contribute to soil compaction, high carbon dioxide levels, and low oxygen saturation. Poor soil aeration is typically caused by soil compaction. The finer the soil, the greater its susceptibility to compaction. The smaller the particles, the more tightly they adhere, leaving less space for oxygen. Neither plants nor aerobic terrestrial biota can thrive in the absence of oxygen.

Regardless of whether it is caused by natural forces or human activity, waterlogging makes soil aeration challenging. After downpours, floods, or heavy irrigation, water fills the pore space of the earth, displacing air and reducing the quantity of oxygen to nearly zero. When land is occupied by water, air cannot permeate the area. When it dries out again, the equilibrium is restored. The air returns to the soil while the water evaporates.

Organic matter improves the fertility of the soil. However, the degradation of organic materials results in a significant carbon dioxide release. Therefore, when the organic matter level is very high, carbon dioxide production can be substantial. As a result, its removal is hindered and can reach dangerous levels. The retention of carbon dioxide also interferes with the delivery of oxygen, which plant roots require and get through the exchange of atmospheric and terrestrial air. To avoid impairing the soil's proper aeration, it is necessary to add organic matter in moderation or in a partially degraded form.

Manure from grazing livestock contributes to soil fertility. Nonetheless, by trotting through the pastures, the livestock compacts the soil with their hooves if they remain in the same location for too long. In this regard, rotating pastures is an efficient solution to the issue.

Farmland can also be compacted by heavy machinery. For this reason, the movement of heavy field equipment, such as fodder harvesters, manure spreaders, etc., should be restricted, and their frequency of usage should be intelligently minimized. The condition is exacerbated when the soil is saturated. Similar problems can be caused by frequently driving vehicles or equipment on your garden land.

Compaction prevents the correct development of plant roots, resulting in long-term production reductions. When compression is combined with dry weather conditions, the situation is dire. In this instance, not only are the roots unable to breathe, but they also cannot absorb water and nutrition. In contrast, when compaction is coupled with humid climate conditions, it promotes soil erosion through the quick flow of liquids.

The purpose of techniques designed to improve soil aeration is to add oxygen to the topsoil so crop roots and soil microbes can access it. Additionally, aeration softens and increases the penetration properties of the top layer. There are multiple techniques for improving soil aeration, each of which depends on the area's size and the soil's qualities.

In raised bed, no dig, or similar systems, compaction is most often not a problem, and these aeration methods will not be needed. If you are starting out with a bad compaction problem, it may be a good idea to implement soil aeration methods before starting a no dig garden. But it will only need to be done once.

If compaction and aeration problems do exist, a radical strategy recommends removing and replacing the land cover. For a small garden, this may be a favorable solution, as you would basically be starting from scratch. However, it is not always possible, not to mention it is expensive and time-consuming to do so for a large area. Standard aeration improvement techniques include solid tine aeration, hollow tine aeration, and liquid aeration.

The solid tine aeration method (also known as the spike method) disturbs the soil the least by creating holes for air to infiltrate. It aerates a relatively narrow area at one time due to limited coverage of applicable tools, such as soil aeration shoes, prongs, rollers, and mower attachments. All of these tools have spikes for penetrating the ground. Aeration shoes, prongs, and rollers require walking or manual operations by pushing or rolling. These options take more physical strength. Attachments for lawnmowers require less human effort because they are attached to the machinery. The solid tine or spike aeration method is suitable for sandy soils. However, it is unsuitable for clay soils due to the spike-like formation which causes compaction.

The hollow tine aeration method (also known as the core method) eliminates "plugs" in the soil, which is particularly important for compacted clay soils. The plugs consist of clay, roots, thatch, etc., in the topsoil layer. This method does not pierce the earth as with the solid tine method, but instead it pulls parts of the earth out of the topsoil, leaving them on the surface. Hollow tine soil aeration equipment includes manual aerators and mower attachments. This form of aeration leaves the field or garden relatively unkempt but offers several advantages. The advantages of aerating the soil with hollow tines are:

- Improved gaseous exchange
- Increased oxygen saturation in the root zone
- Raised water infiltration
- Enhancement of the soil's structure
- Integration of organic matter

The solid tine and hollow tine methods operate directly with the soil, whereas liquid solutions contribute to aeration differently. The components of liquid aerators are a wetting agent and nourishment for soil-dwelling biota. Wetting agents enhance infiltration, allowing water to penetrate deeper into the soil. In addition, they enable bacteria to burrow deeper. By digging, they improve soil aeration and allow water to percolate deeper, which promotes root growth.

Liquid aerators also contain bacteria food (mainly algal extract) to stimulate their activity. This is advantageous for earthworms that travel underground, as it increases the soil's porosity. It allows more significant air and water penetration through the "lanes" produced. Additionally, earthworms and insects decompose organic materials, hence boosting fertility.

Methods of soil aeration are most effective when combined. For instance, it is advantageous to use liquid aerators one week before hollow tine aeration.

Tillage is not the most effective method for aerating soil. Tillage is the most intense form of land disturbance since it entails excavating, churning, and chopping the soil into smaller pieces. Tilling operations minimize compaction and increase oxygen availability, but their benefits are only temporary. The negative impacts of this frequent significant soil disturbance make it a problematic practice. Negative impacts of tillage include:

- Soil health disturbance due to risk of water and wind erosion
- Destruction of beneficial microorganisms
- Uncovering of weed seeds from below ground layers

Unfortunately, tillage is one of the most common methods of soil aeration in commercial farming operations, although the use of no-till methods is increasing. In the long term, the most substantial negative impact of tillage is soil erosion.

The availability of nutrients for crops is directly proportional to the soil's aeration level. Poorly aerated and nutrient-deficient soils restrict plant growth, whereas well-aerated soils offer more ideal growing conditions.

Nitrogen-fixing plants are responsible for organic nitrogen fixation and mineralization. Aerobic bacteria convert organic nitrogen into plant-digestible forms, a process that can only occur in the presence of adequate soil aeration. Inadequate aeration drives the transformation of nitrates into nitrous oxide, one of the most potent greenhouse gases. In addition, denitrifying bacteria are more prone to deplete nitrates from crops in poorly aerated soil.

The valence (combining power) of manganese and iron is high in well-aerated soils and low in poorly aerated soils. In well-aerated soils, sulfur is represented as sulfate, which is ideal for plant growth. Sulfate transforms to sulfide under inadequate aeration (waterlogging), and sulfide and hydrogen sulfide damage plant growth.

A nutrient imbalance due to poor aeration can create a variation in root production, which affects the entire plant and ultimately results in yield losses. Thick, short, black roots with odd forms, hairs that are not fully grown, etc. indicate poor aeration.

The increased susceptibility of the crop to infections and root rot fungi is another significant drawback of inadequate soil aeration. Therefore, an improvement in soil aeration is required if a problem is discovered, as it becomes an efficient preventative measure that minimizes the danger of crop and tree diseases.

If there is a problem discovered in your vegetable garden, it would be best to apply soil aeration methods to address the issue, then switch to a no dig gardening system. This will prevent any future issues with soil aeration and have many additional benefits for your vegetable garden.

Soil Microbes

Microbes are microscopic, unicellular organisms that are invisible to the naked eye. They are sometimes called microorganisms or microscopic organisms because they are only visible under a microscope. They comprise over 60 percent of the planet's living substance.

The term "microbes" is used to refer to a variety of life forms with varying sizes and characteristics. These microorganisms consist of:

- Bacteria
- Fungi
- Protists
- Viruses
- Archaea

Microbes can be both beneficial and dangerous. Certain bacteria can contaminate food and other items and cause severe infections and disorders. Others have an essential part in sustaining environmental equilibrium. Microbes are everywhere, including on the objects we touch, in the air we breathe, and within our bodies. To be able to coexist with all of these microorganisms, humans employ a variety of methods.

On the one hand, there are numerous microorganisms that we attempt to avoid. Consider the foodborne illness-causing bacteria Listeria monocytogenes and Salmonella enterica. We utilize refrigeration to inhibit the growth of harmful microorganisms and make food safe for consumption. On the other hand, certain microorganisms perform practical activities, and we grow them on purpose. For instance, bakers employ warm temperatures to encourage the growth of the yeast Saccharomyces cerevisiae to cause bread dough to rise.

Some microorganisms can be both harmful and beneficial, depending on the circumstances. For instance, Escherichia coli can cause gastrointestinal sickness when consumed, yet it can make synthetic insulin that saves lives in an industrial context. Understanding the intriguing growth of microbes enables us to design methods for maintaining a balance with these species.

Even though many microbes have unique features and capabilities, they share a few commonalities. The majority of microorganisms consist of one or a few cells. Each microbial cell has a cell membrane. The membrane regulates the passage of substances into and out of the cell. This helps the cell take in nutrients while eliminating trash. Some microorganisms also possess a cell wall. The wall offers a structure for enclosing the cell's internal components. Within each cell's core is the DNA encoding its genome. Other facilities in the cell provide crucial metabolic processes for life.

In contrast to a rise in cell size, microbial growth usually refers to an increase in the number of cells. Numerous microorganisms are unicellular, consisting of a single cell. The size of any unicellular microorganism is constrained by the capacity of the cell's critical components to ensure its existence. For example, when cells get excessively large, the integrity of the cell wall is compromised. The solution to cell size limitations is cell division or producing other cells from the parent cell. Consequently, the population increases even though the size of each member of the population remains fixed.

During a typical growth cycle, a single-celled microorganism divides into two identical daughter cells. The original cell, also known as the parent cell, duplicates its DNA and produces enough material to construct the membrane, wall, and molecular machinery for two cells. The parent cell expands somewhat to accommodate these additional components. The center of the parent cell begins to contract, and a new section of the cell wall is assembled at the location of contraction. This process continues until the parent cell divides into two cells with fully developed cell walls.

The cells that result are known as daughter cells. Because both daughter cells are identical, cell division is also known as replication. This type of replication results in a rapid rise in the number of cells, as each daughter cell repeats the cycle by becoming a parent cell. Differently shaped microorganisms may undergo cell division somewhat differently, but the fundamental principles remain the same.

So long as the conditions are suitable, a single cell will continuously make two new cells. Every cycle results in a doubling of the population's cell count. The term for this is exponential growth. The division cycle of E. coli might be as brief as 20 minutes under certain conditions. This rapid division results in the rapid growth of the population.

Eventually, the population will be large enough to have detectable effects, such as the creation of physical structures. This could be a "colony" of bacteria on solid growth media in a laboratory. It takes over a million microbes to build a visible structure, but under perfect conditions for E. coli, this can occur in as little as eight hours.

Certain single-celled microorganisms generate new cells asymmetrically. In this condition, a single daughter cell is produced by a process known as budding. The parent cell causes a tiny protrusion known as the bud during the budding process. The materials required to sustain a new cell are given to the bud, which will eventually separate from the parent cell to generate a new daughter cell. The parent cell continues to produce buds, but the daughter cells formed do not divide.

There are also multicellular microorganisms, such as algae and mold-forming fungus. Multiple cells in these microorganisms collaborate to keep the organism alive. Each cell may serve slightly distinct tasks inside the organism. When a parent cell divides, each daughter cell begins to execute specific duties in response to its environment. As new cells emerge and acquire new activities, the whole organism expands. This is comparable to the growth of larger species, such as mammals.

Environmental circumstances have a profound effect on all forms of microbial growth. Nutrients and energy are among the essential elements for microbial growth. Microbes, like animals, require carbs, lipids, proteins, minerals, and vitamins to survive. Utilizing nutrients and transforming them into cellular structures demands energy. Depending on the types of chemicals it is capable of synthesizing, each microorganism has distinct dietary needs. Most microorganisms are relatively robust and can thrive under various nutrient conditions. However, microorganisms develop more slowly when nutrients are scarce.

Temperature also influences microbial development. The optimal growth temperature for the majority of microorganisms is determined by the ability of proteins within the cell to function. At low temperatures, microorganisms develop more slowly. At higher temperatures, microorganisms multiply more rapidly. Typically, pathogens grow most quickly at average body temperature but at lower temperatures outside the body or when the body temperature rises during a fever.

Extremely high temperatures denature the components necessary for cell survival and are fatal to various microorganisms. Nevertheless, a few unusual microorganisms prefer to grow at extremely high or low temperatures. These microorganisms, known as extremophiles, can grow near hydrothermal vents with temperatures above boiling or in the presence of solid ice.

Even when nutrients are present, and the temperature is optimal, numerous additional environmental conditions might affect microbial development. These factors include acidity, water availability, and air pressure. Each microbe prefers a unique combination of ecological characteristics. In general, each microbe thrives optimally under particular conditions and poorly under others. Specific growth preferences are as varied as the varieties of microorganisms.

Decades of research have led to the current understanding of microbial development and the above principles. Establishing common principles allows us to target large groups of microorganisms, while individual development requirements enable us to target specific microorganisms. This information permits the regulation of microbial development, which facilitates many of our current interactions with bacteria.

Numerous control methods aim to eradicate hazardous microorganisms from food or equipment. For example, high temperatures are frequently utilized to eliminate microorganisms during cooking and pasteurization. In this manner, potentially dangerous organisms are eradicated from the food product, rendering it safe for consumption and storage.

Similarly, the chemicals in disinfectants can harm or kill microorganisms on a wide range of surfaces. Alcohols such as ethanol and isopropyl harm the membranes of cells. Without this protective shell, microorganisms cannot regulate what enters and leaves the cell. Therefore, microorganisms are unable to retain vital nutrients and water. Alternatively, hydrogen peroxide damages cellular structures. As hydrogen peroxide decomposes, it generates chemicals known as free radicals, destructive to proteins and DNA. Soaps are also utilized to eliminate microorganisms from surfaces physically.

Microbes are displaced by the chemical qualities of soaps and the physical force produced when wiping a surface.

When bacteria cannot be eradicated from a substance, such as food goods that cannot be cooked to high temperatures, it is possible to use methods to inhibit their growth. Recognizing the impact of temperature on growth supports the necessity of refrigeration. As previously stated, cold temperatures inhibit the growth of microorganisms; therefore, refrigeration can postpone the growth of microorganisms in certain food products.

In contrast, when we wish to take advantage of bacteria, we attempt to maximize their development conditions. This is why dough with active yeast is left at a warm temperature to promote rapid yeast growth. If the dough is chilled, it will rise considerably more slowly. Similarly, to use E. coli for insulin production, we provide the bacteria with growth-promoting nutrients.

There are a vast variety of microbes in the soil. There are more microbes in one teaspoon of healthy soil than there are people on earth. Some soil microbes are beneficial, and some are harmful. The microbes with negative effects can cause plant diseases such as various forms of blight. The microbes with positive effects perform important roles within the soil, including:

- Nitrogen-fixation
- Phosphorus solubilization
- Transform nutrients into mineral forms that plants can use

- Decompose organic residues and recycle soil nutrients
- Suppression of pests and pathogens
- Improvement of plant stress
- Decomposition that leads to soil aggregation

Specifically, various types of microbes provide the following roles in the soil:

- Bacteria is the crucial workforce of soils. They are the final stage of breaking down nutrients and releasing them to the root zone for the plant.
- Actinomycetes were once classified as fungi, and act similarly in the soil. However, some actinomycetes are predators and will harm the plant while others living in the soil can act as antibiotics for the plant.
- Like bacteria, fungi also lives in the rootzone and helps make nutrients available to plants. For example, Mycorrhizae is a fungi that facilitates water and nutrient uptake by the roots of plants to provide sugars, amino acids and other nutrients.
- Protozoa are larger microbes that love to consume and be surrounded by bacteria. Nutrients that are eaten by bacteria are released when protozoa in turn eat the bacteria.
- Nematodes are microscopic worms that live around or inside the plant. Some nematodes are predators while others are beneficial, eating pathogenic nematodes and secreting nutrients to the plant.

To encourage beneficial microorganisms in your vegetable garden:

- Add compost and other organic materials regularly.
- Sow cover crops between planting cycles.
- Keep your soil well-watered.
- Avoid physical disturbances.
- Mulch your beds.
- Avoid pesticides, herbicides, and fungicides.

Soil pH

The pH of the soil is a measurement of its alkalinity or acidity. pH is measured on a scale from 1 to 14, with 7 representing neutrality. For optimal plant growth, it is necessary to maintain the proper pH.

A pH value measures the concentration of hydrogen ions. Technically speaking, the term pH refers to converting the concentration of hydrogen ions, which typically ranges from 1 to 1,014 gram-equivalents per liter, into numbers between 0 and 14. The concentration of hydrogen ions in neutral (neither acidic nor alkaline) pure water is 107 gram-equivalents per liter, corresponding to a pH of 7. A solution with a pH below 7 is considered acidic, whereas a solution with a pH over 7 is considered basic or alkaline.

Most plants prefer a pH between 6.5 and 7.5, considered neutral. However, many plants have more particular pH requirements, such as blueberries and azaleas, which prefer an acidic soil, and lilacs and clematis, which favor an alkaline environment. There are also plants, such as hydrangeas, whose blossom color varies depending on the pH of the soil.

Generally speaking, if your plants are thriving and showing no indications of distress, the pH of your soil is likely within an acceptable range. However, if your plants are stressed, discolored, or not producing, you should conduct a pH test.

Typically, the pH is measured with a pH meter, which converts the difference in electromotive force (electrical potential or voltage) between electrodes placed in the solution to be tested into pH readings. The meter's readout may be digital or analog (scale and deflected needle). Digital readouts provide the advantage of precision, whereas analog readouts provide more accurate indications of change rates.

There is widespread usage of battery-powered portable pH meters for testing the pH of soils in the field and in home vegetable gardens. Less accurately, pH tests can also be conducted with litmus paper or by mixing indicator dyes in liquid suspensions and comparing the resulting colors to a pH-calibrated color chart.

If you don't want to purchase a pH meter, you have several alternatives for evaluating the pH of your soil. Many inexpensive do-it-yourself testing kits can be found in local garden centers. Most do an excellent job of indicating which end of the scale your soil falls on, which may be sufficient for making modifications. You can send a soil sample to a lab or bring it to your local Cooperative Extension office for more precise measurement. They will charge a modest price, but you will have a more accurate understanding of the state of your soil.

Changing soil pH takes time, typically months, and must be a continuing process. The soil will ultimately return to its normal pH if left alone. Unless your soil is excessively acidic or alkaline, you will not need to modify it all. You can change only the regions where plants with varied pH requirements are grown.

The pH of acidic soil is generally increased by adding lime or dolomite. The pH of alkaline soil is generally decreased by adding sulfur. How much to add depends on the soil's current pH, its texture (clay, sand, humus), and what will be grown. Your soil test kit or the report you receive from the lab or extension service should outline the necessary steps. Otherwise, follow the instructions on the product you are using.

Soils having a high concentration of organic matter and clay will be more resistant to pH changes and will therefore require more treatment than soils low in organic matter and clay. While soil pH being acidic indicates the need for lime, and a soil pH being alkaline indicates the need for sulfur, it is not a good indicator of how much lime or sulfur is necessary.

To maintain a soil pH within an acceptable range, it will be necessary to reapply lime or sulfur annually or bi-annually. This is best accomplished in the fall or off-season so that the amendment has time to permeate the soil and not injure plant roots. In addition, you should retest your soil every three years to determine if any alterations are necessary.

Most soils have pH values between 3.5 and 10. In regions with more significant precipitation, the natural pH of soils runs from 5 to 7, but in drier areas, the range is from 6.5 to 9. Soils can be categorized based on their pH value as follows:

- Below 4.0 – Acid sulfate soil

- Below 5.5 – Highly acidic soil
- Below 6.5 – Acidic soil
- 6.5 – 7.5 – Neutral soil
- Greater than 7.5 – Alkaline soil

The soil's natural pH is determined by the rock it developed from (parent material) and the weathering processes that operated on it, such as climate, vegetation, terrain, and time.

Soil pH influences the solubility of nutrients and compounds in soil water and, consequently, the availability of nutrients to plants. Some nutrients are more accessible in acidic environments, whereas others are more accessible in alkaline environments. However, most mineral nutrients are readily available to plants when the pH of the soil is close to neutral.

The formation of highly acidic soils (pH below 5.5) might result in poor plant growth as a result of one or more of the following:

- Aluminum toxicity
- Toxicity of manganese
- Calcium deficiency
- Magnesium deficiency
- Low concentrations of essential plant nutrients, including phosphorus and molybdenum.

Alkaline soils may have nutrient deficiencies with zinc, copper, boron, and manganese. Soils with a pH greater than 9 are also likely to contain high salt concentrations.

For most plants, the optimal pH range for soil is between 6.0 to 7.5. It is essential to monitor pH levels regularly if there are any crop issues.

Soil Salinity

Saline soils are not optimal for crop growth. While some salinity in soil is normal and necessary, it becomes a problem when enough salts accumulate in the root zone to negatively affect plant growth. If you are having your soil tested, most labs will include a salinity measurement.

In extremely salty regions, frequent localized irrigation is required to remove salts from the root zone of plants. Soils can become salinized over time if they receive water with a high concentration of dissolved salts and if drainage is poor. This is especially prevalent in irrigated areas of arid and semi-arid regions.

Macronutrients and Micronutrients

Primary macronutrients, secondary macronutrients, and critical microelements are all essential to plant nutrition and must be present in the soil for healthy plant growth. In addition, these components must be present in the correct proportions.

Large quantities of primary macronutrients are consumed by plants and their consumption of secondary macronutrients is lower. Each of these nutrients serves a unique purpose in sustaining the plant. Each nutrient deficiency has distinct negative impacts on the plant's overall health, depending on which nutrient is lacking and what quantity.

Plants only require light, water, and about 20 elements to support all of their biochemical needs. Three of those 20 elements are removed from the air as carbon dioxide or water. These three elements are carbon, hydrogen, and oxygen. The remaining 17 elements are collected from the soil. These are separated into macronutrients and micronutrients necessary for plant growth.

Macronutrients are necessary for plant development and overall health and are consumed plants in large quantities. Primary macronutrients are consumed in the largest quantity and include Nitrogen (N), Phosphorus (P), and Potassium (K).

Nitrogen (N) is necessary for plant growth because it plays a crucial role in energy metabolism and protein synthesis. The plant absorbs nitrogen in the form of nitrate. This macronutrient is intimately linked to plant development. It is essential for photosynthesis and the creation of chlorophyll. Nitrogen is primarily engaged in the aerial zone or the visible portion of the plant. It encourages cell multiplication. A nitrogen deficit causes a decline in vitality and color. Beginning at the bottom of the plant, growth slows, and leaves start to drop.

Phosphorus (P) is necessary for root development, which it accelerates. In the aerial zone, phosphorus promotes flowering. Although phosphorus is required during the plant's growth stage, it plays a considerably more significant role during the flowering stage. Phosphorus is essential for the delivery and storage of energy. It enhances the plant's general health and increases its resistance to unfavorable climatological circumstances. Phosphorus is required for the creation of organic compounds and the proper execution of photosynthesis. Phosphorus deficiency causes late, poor flowering, leaf browning and wrinkling, and an overall lack of vitality.

Potassium (K) is involved in the management of water and the transfer of reserve compounds in plants. It boosts the ability for photosynthesis, strengthens cell tissue, and stimulates nitrate absorption. Potassium enhances blooming, glucose synthesis and enzyme production. This, in turn, increases the plant's resistance to adverse conditions, such as low temperatures and keeps it from withering. A potassium deficiency diminishes a plant's resistance to dry conditions, frost, and fungal assault. This results in an imbalance of other nutrients, such as calcium, magnesium, and nitrogen. When there is insufficient potassium, dark patches emerge on the leaves.

Secondary macronutrients are also necessary for plant nutrition. They are consumed by plants in less quantity than primary macronutrients but in more quantity than micronutrients. The secondary macronutrients are Calcium (Ca), Magnesium (Mg), and Sulfur (S).

Calcium binds to the cell walls of plant tissues, thereby maintaining the cell wall and promoting cell wall development. Calcium is also essential in cell development and growth. It enhances plant vitality by promoting root development and growth. Calcium aids in the retention of minerals in the soil and the movement of these minerals. It neutralizes harmful chemicals in plants and aids in the development of seeds. Calcium regulates and stabilizes numerous processes within the plant. A calcium deficiency results in yellow and brown patches on the leaves. Additionally, it decreases plant development in general.

Magnesium forms the chlorophyll molecule's nucleus and is, therefore, needed for photosynthesis. This makes it an essential ingredient for plant growth. Magnesium facilitates phosphorus absorption and transport. It contributes to the plant's ability to store carbohydrates. Magnesium is an enzyme activator, activating more enzymes than any other mineral. Magnesium deficiency causes weak stalks, a loss of green in the oldest leaves, and the formation of yellow and brown scars, even though the veins of the leaves stay green.

Sulfur contributes to the synthesis of chlorophyll. It is required for photosynthesis and involves protein synthesis and tissue development. Sulfur is essential for nitrogen metabolism and increases nitrogen efficiency. Sulfur also strengthens plant defenses in general. When a sulfur deficiency occurs, which is uncommon, the plant becomes lighter in color, taking on a pale green hue. Similar to a nitrogen deficit, general chlorosis might be observed.

A micronutrient is a mineral, vitamin, or other substance that is essential, even in very small quantities, for plant growth or metabolism. The 11 plant micronutrients include Boron (B), Zinc (Zn), Manganese (Mn), Iron (Fe), Copper (Cu), Molybdenum (Mo), Chlorine (Cl), Nickel (Ni), Cobalt (Co), Sodium (S), and Silicon (Si).

Nutritional Monitoring of Crops

A soil test is the only way to check the levels of required plant macronutrients and micronutrients. The soil's pH and organic matter content can be tested at the same time. Testing your soil at least once every three years can help prevent mistakes such as applying too much or the wrong kind of fertilizer or amendment, which increases expenses and negatively impacts the soil over time.

There are three steps to the soil test process:

- The collection of a sample
- Laboratory examination
- The creation of amendment instructions for the desired crop

The soil test process will help you determine how much of each required plant nutrient is present in the soil and should recommend amendments to address any deficiencies found.

Preferably, soil samples should be collected after the soil is prepared for the season but before fertilizer application, a few weeks before planting. This allows time for the results to be considered and addressed.

Fifteen to 20 subsamples should be obtained from various sections of each garden plot. These should then be thoroughly combined to create a two pound composite sample. To obtain a representative sample, avoid the garden's edges. If a garden plot contains many soil types, different prior harvests, or is exceptionally vast, it is recommended to treat it as if it were two or more plots, that is, to collect 15 to 20 samples from each section and to create two or more composite pieces. Soil samples are typically collected from the top 10 to 12 inches of the active soil layer for annual crops. Soil samples should be taken from the depth where the crop grown will take root.

    Immediate labeling of soil samples with the producer's name, plot name or number, date, and sample depth is crucial. In addition to these statistics, the laboratory will need to know the desired crops to offer the most appropriate recommendation for amendments. They can also make general recommendations for varied vegetable gardens.

No Dig Gardening

No dig or no till gardening is the best way to garden in the ground while regenerating and maintaining the soil's health and fertility over time. The no dig method allows you to create a large vegetable garden at a more affordable price than constructing raised beds and has the benefit of restoring your native soil into a productive ecosystem.

The no dig method, as its name implies, leaves the soil undisturbed. The only soil disturbance within this system is the shallow cultivation necessary to remove an occasional weed or to plant seeds and seedlings. You feed soil life with organic matter on the surface, as it happens in nature. This adds fertility while maintaining drainage and aeration. No dig works on all soil types including heavy clay.

No dig gardening is advantageous for a variety of reasons:

- Increased productivity compared to traditional garden methods where the ground is tilled.
- No need to own a rototiller.
- Fewer weeds compared to traditional methods. Regular applications of compost smother most weed seeds and young seedlings and provide a rich source of nutrients. The few weeds that grow are easy to remove by hand or with a shallow, fast-moving hoe.
- Less time required to prepare beds for planting.

- Millions of beneficial microbes within the compost break organic matter down into natural plant food at a rate that the garden plants can use, eliminating nutrient runoff.
- These same organisms also contribute to long-term soil improvement as they build humus, the basic organic soil component that promotes good air and water movement through the soil and is linked to reductions in plant diseases.
- The soil's structure, health and fertility is improved over time, instead of being degraded over time as with traditional methods where the ground is tilled.

How to Start a No Dig Garden

There are no special skills or tools required to start a no dig garden. It is much less labor intensive than a traditional tilled garden since operation of a rototiller is eliminated from the process.

When locating your no dig garden, the following conditions should be met:

- Most veggies require 6 to 8 hours of direct sunshine each day (sometimes known as "full sun"), particularly after lunch.
- You ideally want a location with level, even terrain.
- Consider convenient access from the house for weeding and harvesting.
- Avoid placement in a frost pocket or a windy area.
- Because soil needs to drain efficiently, stay away from marshy or damp places.

Follow these steps to start a no dig garden:
1. Prepare the ground.
    a. If you're creating a no dig garden over bare soil or an existing garden bed, skip this step.
    b. Remove any coarse, woody weeds like briars, vines, and shrubs.
    c. If you're on rocky ground, add a three to four inch layer of sticks, branches, and dry leaves, which will aid in drainage.

d. If you're over grass, mow the grass very low, then, optionally, fertilize with bloodmeal and bonemeal and water that in. This helps rot the grass more quickly.
   e. Optional: Apply soil aeration improvement methods if major compaction issues are observed.
2. Lay down newspaper or cardboard.
   a. Lay down a ¼ inch thick layer of newspaper, overlapping edges by six inches, and water it down. Do not use glossy print paper (inserts/ads) as they contain toxic inks.
   b. Or, cover the area with a double layer of cardboard, overlapping edges by six inches, soaking each layer with water. Do not use coated cardboard boxes as they contain toxic inks.
3. Lay down hay or straw.
   a. Add a four inch thick layer of hay or straw.
4. Add manure or compost.
   a. Spread a two inch layer of aged manure or high quality compost over the straw.
   b. Optionally, add worm castings.
   c. Water the compost in.
5. Lay down more hay or straw.
   a. Add another three to four inch thick layer of hay or straw.
6. Add more manure or compost.
   a. Spread another two inch layer of aged manure or high quality compost over the straw.
   b. Water the compost in.
7. Lay down more hay or straw.

　　　　a. Add another three to four inch thick layer of hay or straw.
　8. Plant it up!

　You can start planting in your no dig garden right away. Gather your plants and make a hole in the top straw layer with your hands, just big enough for the plant's root ball (about four to six inches wide). Fill the hole with compost, then plant seeds, seedlings, or small plants, and water in well. You can plant anything in a no dig garden that you would plant in a traditional vegetable garden. Follow the spacing instructions on the seed packet or for the specific seedlings you are planting out.

Raised Garden Beds

At the home garden scale, or even at the urban/small scale market garden scale, raised bed gardening can be ideal because it allows you to customize your soil and ensure that it is of the highest quality. It is also a lower maintenance gardening strategy since there is less weeding required than with in the ground gardening methods. Raised beds can be perfect for busy families looking to grow some of their own food. No special DIY abilities are needed to construct a raised garden bed.

A raised bed is a freestanding box or frame filled with high-quality soil and typically has no bottom or top. The bottom of raised beds is typically open so that plant roots can spread and access soil nutrients below ground level.

Raised beds are advantageous for a variety of reasons:

- They drain effectively and reduce erosion.
- Because the soil is lifted above the ground, they warm up earlier in the spring and extend your growing season.
- Planting intensively is made feasible by raised beds because you have control over the soil you put in them. Plants grown closely together in raised beds also mature more quickly.
- Since you aren't walking in the bed, the soil doesn't get compacted and stays loose.

- Raised beds prevent weeds from taking over because the beds are elevated away from surrounding weeds and packed with disease- and weed-free soil.
- Raised beds aid in maintaining order and control.
- Less bending and kneeling results in more accessible and pleasant gardening tasks.
- Raised beds make it easier to separate and rotate crops each year.
- Raised beds enable easier square-foot gardening and companion planting.
- And finally, raised beds are attractive.

You can construct your raised beds with a variety of materials, including wood, metal, stones, bricks, and cement blocks. It is best to avoid pressure-treated or painted wood because these materials may release chemicals or lead into your soil.

Of the natural wood choices, pine is the least expensive. Being a softwood, untreated pine will usually last 3 – 7 years depending on the climate it is exposed to. Hemlock has a little longer lifespan. Woods resistant to rot, such as cedar, redwood, or locust, will last longer but are much more expensive. The best option is cedar since it resists rot and is strong, lasting 15 – 20 years. The natural oils in the wood make it insect resistant as well. However, many people opt for pine instead due to the affordability.

A relatively new option is "recycled plastic lumber". It will last forever and is environmentally sustainable because it is made from recycled water bottles and similar plastic containers. However, it currently comes with a hefty price tag. Composite wood is another recent product from wood fibers and recycled plastic. It is durable and resistant to rot but also highly expensive. The price tag may be justified since these materials are said to last at least 50 years.

As long as you know where they came from, pallets can be an affordable source of materials for garden beds. Avoid using pallets previously treated with methyl bromide, a known endocrine disruptor that can harm your reproductive system. Many pallet manufacturers ceased using the chemical in 2005, although there are still a lot of outdated pallets around. On the pallet, search for the letters "HT," which stand for "heat treated." Use caution when using a pallet in your garden if there is no stamp, or an HT cannot be verified on the surface.

Brick or concrete can be used to construct raised beds. However, keep in mind that concrete will gradually raise the pH of the soil, necessitating soil amendments. Mediterranean-style herbs like rosemary and lavender thrive in the additional heat collected from concrete blocks. Strawberries or herbs can be inserted into their holes and then covered with soil. The concrete blocks (or "cinder" blocks) are 16 inches long by 8 inches high and are usually affordable at big box retailers. Rocks and stones are abundant in some regions and can make good free edging for raised beds.

The ideal width for a raised garden bed is four feet. This allows you to reach the garden without stepping into the bed. If you have to step in your beds, the soil becomes more compacted, making it more difficult for plant roots to absorb oxygen. It will be challenging to reach the middle of an extensive bed for weeding and harvesting. If your raised bed will be constructed against a wall or a fence, it is better to keep it no wider than two to three feet, since you'll only be able to get to the garden from one side.

The length of the raised bed is not as crucial. A 4'x4' bed, 4'x8' bed, 4x12' bed, or a 4'x16' bed can be made to coincide with common dimensional lumber lengths. Although you can make beds as long as you want, building several shorter beds may be more accessible than making one long bed. Additionally, separate beds are best for many crop families.

Typically, lumber will be available in a standard size that is 6 inches tall. Common dimensions are 2 inches by 6 inches by 8 feet. The boards can be stacked. Two "2 x 6" boards can be stacked to 12 inches. You can also purchase 2" x 12" boards to avoid having to stack and join the boards together, but this is more expensive.

Of course, you can go higher than 12 inches (18 inches, 24 inches, or 36 inches), but remember that the increased soil's weight will put more pressure on the sides. Any bed that is taller than 12 inches must have cross-supports.

Think about what you want to grow and how much soil depth the crop requires underground. Crops with deep roots, such as tomatoes, squash, potatoes, carrots, and parsnips require a soil depth of 12 to

18 inches. The roots of these plants cannot penetrate far enough to reach nutrients if the soil is not loose enough. Crops with shallow roots, such as lettuce, greens, and onions, require a minimum of 6 inches of soil depth.

You may want to make sure your beds are between 12 and 18 inches deep to be safe. Unless the soil under your raised bed is particularly rocky or has a heavy clay content, a 12 inch high raised bed with an open bottom should be sufficient. Depending on the height you choose for your bed, you may want to loosen the dirt beneath the ground to the appropriate degree. For instance, if you want to plant root vegetables in a bed that is 6 inches high, you may want to loosen the soil there by another 6 to 12 inches. There would be no need to do so if you were planting crops with shallow roots.

When locating raised garden beds, the following conditions should be met:

- Most veggies require 6 to 8 hours of direct sunshine each day (sometimes known as "full sun"), particularly after lunch.
- You ideally want a location with level, even terrain.
- Consider convenient access from the house for weeding and harvesting.
- Avoid placement in a frost pocket or a windy area.
- Because soil needs to drain efficiently, stay away from marshy or damp places.

When getting the site ready, run strings between stakes or sticks to outline the area you want your raised bed to go in. Keep it around four feet wide, as previously suggested, to provide easy access to the center. If sod or grass is in place, cut it extremely short. Then you may choose to dig it up, keeping the clumps to one side. A manual "kick" sod cutter will make the job easier. Alternatively, after cutting the grass as short as possible, you can cover it with a thick, high quality landscape fabric, or with several cardboard boxes laid in an overlapping fashion.

Some gardeners choose not to remove the turf because, provided the soil is thick enough, it will smother out the grass and weeds growing there. This is similar to the no dig technique. The theory is that weeds grow more quickly and require more weeding because digging exposes more weed seeds to the soil's surface. Digging also rips up the intricate life and fabric of your soil, making it less able to drain correctly and retain moisture, which means you'll need to feed plants more frequently.

Here is a "no dig" raised garden bed site preparation guide:

1. Trim the weeds and grass as near the ground as close as possible. Then, use cardboard or a thick layer of newspaper to cover the area, which will eventually rot into the soil, and smother the grass and weeds. (Be sure to remove any tape and staples, as they won't break down.)

2. To prevent weeds from slipping through cracks, be sure to overlap the cardboard/newspaper by about six inches. Any source of sunlight will be sought by grass seedlings.
3. Spread a substantial layer of compost over the cardboard (3 to 4 inches).
4. Fill your raised garden bed as normal.
5. Through the actions of worms and other organisms, the compost you place on top of the cardboard should eventually bind to the soil below.
6. Topping off the beds with organic material each year will progressively increase the soil's fertility and health, including the ground below the raised bed's level. This means that growing vegetables with deeper roots should be no problem.

How to Build a Raised Bed with Wood

Building a raised bed out of wood requires only the most basic DIY abilities and can be handled by a complete beginner, because you are just building a box, similar to a sandbox. You could always purchase a raised garden bed kit from a store, but their costs are much higher than building one on your own.

After selecting and preparing your site as directed above, you can follow these steps to construct a 4'x8' raised bed out of wood (or customize the bed to your preferred length).

Tools required to build a raised garden bed out of wood:

- Screwdriver, power drill and bits
- If cutting the boards yourself (as opposed to purchasing them cut from a lumberyard or big box retailer): tape measure and a hand saw or power circular saw

Materials required to build a 4'x8' raised garden bed out of wood:

- Six 8-foot-long pieces of 2"x6" lumber are needed to make a 4x8-foot bed that is 12 inches deep. If having lumber cut, have two of the 8-foot boards cut in half to 4-foot sections for the sides. (Alternatively, three 8-foot long pieces of 2"x12" lumber can be used to avoid stacking the boards. In this case, have one of the boards cut in half to 4-foot sections for the sides.)

- Exterior deck screws 4" – 4.5" long
- Four 12-inch long pieces of 4"X4" lumber for the corners This is optional if using 12" wide boards but still recommended. (Note: The 4"x4" lumber usually comes in standard 8-foot lengths. If you have it cut into eight 12-inch long pieces, you will have enough for two raised garden beds.)

How to build a 4'x8' raised garden bed out of wood:

1. Measure and cut the boards to the specified lengths if they were not already cut at the lumber yard.
2. Decking screws will be used to assemble the boards. It's enough to have two holes at each board's end (or four holes if using the 12" wide boards). Use a drill bit smaller than the screws to create pilot holes. Each board's ends will overlap and be screwed together directly to make the box, so place your pilot holes accordingly. Set the boards down in place. Each board must be positioned so that it overlaps the one before it, with the pilot holes placed at the overlapping ends. Use your power drill to connect the bed together with the exterior decking screws. It is easiest if you have someone holding the boards as you fasten the corners.
3. Once the box frame is constructed, place the 12-inch long pieces of 4"x4" lumber in the corners. Drill pilot holes then screw the boards to these pieces for bracing. These pieces tie the bed together in one piece if you are stacking 2"x6" boards. If you are using 2"x12" boards, the corner braces are optional, but still

recommended to provide more support and a more durable frame.

Filling a Raised Bed

The soil is the most important component of any garden. Organic matter is a crucial component of the soil, helping plants flourish due to the feeding of soil bacteria and easy access to water and oxygen for root systems. The appropriate balance for a raised bed is as follows:

- 40% compost
- 40% topsoil
- 10% moisture retention
- 10% aeration

Compost should make up about 40% of the raised bed filling. Compost is rich in plant nutrients. It can be made at home, but you may have a hard time making enough quantity to initially fill your raised beds (although homemade compost will be ideal to top off your beds with when needed). For the initial fill, compost can be purchased in bulk from a local nursery, which is usually the most affordable method, especially if you are filling multiple raised beds. It can also be bought in bags from a local garden center or big box store. Fresh manure cannot be used directly in your garden; however, you can use aged livestock manure to offset the required amount of compost.

Topsoil should make up about 40% of the raised bed filling. Again, the most affordable source of bulk topsoil is usually a local nursery. They should be able to mix the compost and topsoil together for you at no additional cost, which saves you the labor. They should also be able to deliver it to you for a fee. Topsoil is also available in bags from a local garden center or big box store. We are not talking about potting soil, which is designed for container gardening. Potting soil is too fluffy for raised beds and much more expensive than topsoil. If you are buying it by the bag, the bags should be labeled as topsoil.

Materials that aid in moisture retention should make up about 10% of the raised bed filling. Ideal options are peat moss or coconut coir. Peat moss is usually available in large bags at local garden centers or big box stores and is the most affordable option. There are environmental concerns related to peat moss harvesting. Additionally, peat moss can acidify the soil slightly.

A more expensive but environmentally friendly option is coconut coir. Coconut coir is more of a specialty product (although it is definitely becoming more popular) and it may or may not be available in retail stores near you. You can order coconut coir in blocks to be rehydrated at home, which reduces the shipping cost.

Coconut coir is a natural fiber that revolutionized growing media in the horticultural industry. It has the same beneficial effects of peat moss without the environmental concerns related to its harvesting. Coconut coir is a byproduct of the harvesting of coconuts, and it is reducing the quantity of garbage produced. Previously, all components—from the husk to the inner shell—were thrown away carelessly. The brown and white fibers known as coconut coir found between a coconut seed's shell and outer coating have a wide range of uses in horticultural products.

An in-between option is to use half peat moss and half coconut coir. This would also minimize any acidity caused by the peat moss. If you can afford it, using all coconut coir instead of peat moss as your moisture retention material would be preferred for environmental reasons related to concerns with peat harvesting.

Finally, materials that aid in aeration should make up about 10% of the raised bed filling. Ideal options are perlite or vermiculite. Perlite is the little white balls that look like Styrofoam which you are probably used to seeing in potting mix. It is a naturally occurring mineral. In nature it exists as a type of volcanic glass. Vermiculite is also a naturally occurring mineral. It is made from compressed dry flakes of a silicate material.

Compare local prices to online prices when sourcing perlite or vermiculite. Even at big box stores, they often will only come in small bags. If that is the case in your area, you will save a lot of money by ordering in bulk online. Look for Grade 3 particles between 3-6 mm in size (it should be labeled as vermiculite or perlite for gardening purposes, as finer particles are sold for other purposes but won't work well for gardening). A more affordable alternative to perlite and vermiculite is crushed lava rocks, but perlite or vermiculite are preferred due to their superior performance.

To calculate the amount of each fill material needed, first figure out the cubic feet of the bed you are filling. In our example, we built a 4'x8' bed that was 12 inches (1') high. To get the total cubic feet required, we simply multiply 4' x 8' x 1' = 32 cubic feet (1.185 cubic yards).

Multiply by 0.4 (40%) to get the compost and topsoil amounts, and multiply by 0.1 (10%) to get the moisture retention material and aeration material amounts. This gives us 12.8 cubic feet compost (0.474 cubic yards), 12.8 cubic feet topsoil, 3.2 cubic feet moisture retention materials (0.119 cubic yards), and 3.2 cubic feet aeration materials (these numbers can be rounded for purchasing convenience). It may be necessary to plug the cubic feet numbers into an online calculator when making your purchases depending on how the materials are sold to convert between measurements (such as the "cubic feet to cubic yard calculator" used for this example).

When mixing the fill ingredients together, it is easiest to layer the ingredients like lasagna and mix as you go. Use a hoe, shovel or cultivator to combine all the ingredients. Do it bit by bit until you fill the bed. It will be more difficult if you add the materials all at once.

Once filled, it's also a good idea to keep your raised bed soil mulched with a natural material such as straw, finely shredded leaves, or grass clippings (from grass that was not treated with chemicals and was not seeding when cut). The mulch will help retain moisture for your plants, prevent weed growth, and also serves as a layer of protection for your soil. Natural mulches will also feed your soil with organic materials as they slowly break down. Wood chips do not make the best mulch for vegetable gardens, because as they break down over time they turn the soil into a fungal based system (as you would find in a forest) rather than the bacterial based system that is preferred by most vegetable plants.

When you first start watering your raised bed soil, the materials may settle. If this happens you can simply top off your raised bed with compost. You can also add extra compost to the top of your raised beds in the fall or winter after the growing season has ended. It will help protect the soil and make it more fertile for spring planting. Anytime you notice the soil level of your raised bed getting lower, just top it off with compost. You will never have a problem with erosion, and you will maintain your soil fertility if you do this.

Your soil will be so fertile that you likely won't need to apply plant fertilizer during the first year of use. But as your food crops consume all the nutrients in subsequent years, your soil will require some amendment using compost or a balanced, slow-release fertilizer. If you are regularly adding high quality compost, you may be able to avoid using fertilizers all together, or you may only have to add a small amount in the beginning of the season.

Planting Ideas for Raised Beds

Beginning gardeners may want to try their hand at growing a few of their favorite veggies in one raised bed. A good sized garden can be created from a combined four or five raised beds. Or you can set up ten or twenty raised beds and have a seriously productive homestead.

The depth of your soil, which is equal to the depth of your raised bed plus the depth of the dirt you dug and loosened below ground (if that was needed), is the sole restriction on what you can grow. Leaving the bottom of your raised bed open to the ground will ensure your crops have enough room for their roots to spread.

What thrives in 6" of soil depth?
Basil, chives, cilantro, dill, mint, oregano, parsley, thyme, marigolds, and other annual flowers. Lettuce, salad greens, spinach, onions, leeks, radishes, strawberries

What thrives in 12" of soil depth?
In addition to everything on the 6" list: beans, beets, broccoli, Brussels sprouts, cabbage, cantaloupe, carrots, cauliflower, collard greens, cucumbers, garlic, kale, summer squash, Swiss chard, turnips, lavender, rosemary, sage, borage, calendula, cosmos, lantana, nasturtiums, snapdragons, and sweet alyssum.

What thrives in 18" of soil depth?
In addition to everything on the "6" and "12" lists, you can grow eggplant, okra, peppers, pumpkins, sweet potatoes, tomatoes,

watermelon, and winter squash. These are just examples. Virtually any crop can be grown in 18" soil depth.

Container Gardening

Vegetable gardens in containers are perfect for cramped areas, porches, balconies, and even that tiny patch of your side yard that gets enough sun but isn't big enough for a garden bed. Your potting material is crucial to your container garden's success.

Many types of potting soil and potting mix can be purchased from stores. If you have a small number of containers, you may prefer to purchase soil rather than take the time to mix it yourself. If you do purchase soil for your container garden, be sure it is labeled potting soil or potting mix. It is more expensive than standard garden soil but is mixed specifically to do well in containers.

If you are filling a large number of containers, mixing your own potting soil can save you money. Additionally, because you carefully choose each ingredient in your mixture, you'll never worry about what is in it.

Here is an easy recipe for homemade potting soil for container vegetables:

- Fine Compost (2 parts)
    - Use high quality homemade or locally purchased compost. Passing your compost through a sieve or screen to remove more significant bits from the mix is essential to prepare it for use as potting soil for container-grown veggies.

- Coconut Coir (1 part)
    - This is an environmentally friendly product derived from coconut husk fibers. Coconut coir usually comes in blocks that need to be rehydrated. When used in potting soil for vegetables grown in containers, coconut coir helps retain moisture.
- Perlite or vermiculite (1 part)
    - Perlite is the little white balls that look like Styrofoam which you are probably used to seeing in potting mix. It is a naturally occurring mineral. In nature it exists as a type of volcanic glass. Vermiculite is also a naturally occurring mineral. It is made from compressed dry flakes of a silicate material. Compare local prices to online prices when sourcing perlite or vermiculite. Even at big box stores, they often will only come in small bags. If that is the case in your area, you will save a lot of money by ordering in bulk online if you need more than a couple of small bags. Look for Grade 3 particles between 3-6 mm in size (it should be labeled as vermiculite or perlite for gardening purposes, as finer particles are sold for other purposes but won't work well for gardening). Vermiculite and perlite aid in aerating the soil.
- Worm Castings (about 1 cup)
    - If you conduct your own worm composting or "vermicomposting", you're in the lead. Worm castings are also available to buy. Minerals like potassium, magnesium, phosphorus, and calcium are abundant in worm castings.

- Additional minerals and potting soil additives for container-grown veggies
    - While combining your potting soil batch, you can add organic fertilizer. Alternatively, depending on the veggies you are planting, you might wait and add as necessary. You can add fertilizer all at once to save time. Or you can customize the potting mix to the exact seed or seedling by delaying and adding what you need later.

After mixing your potting soil for container vegetables, use a meter or testing kit to determine the pH. Then recheck it a few days later. You are then prepared to sow vegetable seeds or seedlings in your pots. Potting mix being stored for later use should be kept in a container with a tight-fitting lid to assist in preventing moisture loss.

How to Grow Vegetables in Containers

You don't need to clear a sizable outdoor area or even a backyard to grow some of your own food. You can develop an edible container garden even if your porch or patio is only a small space. Vegetable container gardening offers many opportunities, including the chance to produce and harvest delicious varieties uncommon in supermarkets. Position your containers in a sunny area and select a few different vegetables you and your family will love eating to get started. You'll soon have wholesome, delectable produce blooming just outside your door.

Select your desired containers before you begin planting. Choose containers that will work for the area you have and the veggies you want to grow because the type and size of the container might alter the amount of care your garden needs.

If you're uncertain about the best type of container for growing vegetables, don't worry. Most vegetables don't care much about the container they grow in. The only prerequisites are that the pot has drainage holes so that extra water can flow and that it is big enough to contain the plant.

Due to the porous quality of the material, plants growing in terra-cotta (clay) pots generally require more care while watering than plants growing in other types of pots. Pick a lightweight container if you intend to move your containers around a lot. Containers can get quite heavy after planting, especially after watering. Additionally, consider the color. In the summer, particularly in hot climates, dark colors may make the soil excessively warm for various vegetable crops because they absorb heat. Inversely, darker colors will help the soil absorb heat in cooler climates. Avoid growing veggies in treated wood containers since they can contain chemicals that the vegetables could absorb.

The larger the containers, the better. Larger containers require less watering. Since they contain more soil, larger containers can retain moisture for extended periods. Search for containers with a minimum of 10 inches in width and 12 inches in depth. Also, don't limit yourself to the conventional round flower pot. Half barrels, bushel baskets lined with plastic, and window boxes can all serve the same purpose.

For tomatoes and other deep rooted plants, use larger containers at least 18 inches in depth. Plants with long stems or vines (like tomatoes and cucumbers) will yield more fruit if grown in a container with support. Providing support can be as straightforward as a wire cage placed into the container at planting time. To reduce the chance of plants on trellises tipping over, use larger, heavier containers for these types of plants. Pots designed for small trees are ideal.

Vegetables require potting soil that will allow water to drain efficiently and holes in the container they are in to allow water to drain out. Soil should extend at least two to three inches below the rim of the pots (that extra space at the top will give you room to water deeply without overflowing the container). Immediately before planting, water the soil in.

Consider your favorite foods as a beginning point when choosing plants for your container garden. Most veggies have comparable requirements (full sun and well-drained soil). You can start seeds in your containers, start them indoors and transplant them, or buy plants from a garden center or local nursery, depending on the kinds of vegetables you wish to cultivate.

Check your local Cooperative Extension's planting guidelines for local sowing instructions by type of vegetable (or outside the United States, check the similar organization for your country). Pay attention to the season you are in, as some vegetables are hot weather crops, and some are cool weather crops. For larger vegetable plants, select compact varieties that are marked as "container" varieties when possible.

Example vegetable plants which can be started from seeds placed directly in the container include lettuce, spinach, collards, cucumbers, squash, beans, corn, carrots, and radishes. Adjust each plant's spacing by 3 to 4 inches closer than recommended for standard gardens, and then follow the instructions on the seed package. Plant more seeds than you need because not all of them will germinate, then thin the extra later. However, don't overdo it either or you will waste seeds and spend a lot of time on thinning. Sow about twice as much seed as the number of plants you will need.

Example vegetable plants which must be transplanted or started in the containers indoors include tomatoes, peppers, eggplants, cabbage, broccoli, and cauliflower. (This is the case in most but not all areas, due to the limited outdoor growing season for these crops and the length of time required for plant maturity/harvest). Before placing the root ball in your container for transplanting, gently loosen it by tugging at the roots. You can purchase plastic plant identification tags to aid in identifying each plant.

Before or after planting, mix a balanced organic fertilizer into the soil. Avoid over-fertilizing as this will cause the plants to grow too quickly, making them more prone to topple over, and the flavor won't be as robust. Follow the instructions on the fertilizer's container to begin feeding your vegetables once every two to four weeks, starting approximately a month after planting them. After planting, water the seeds or transplants thoroughly but gently to help them settle. Mulch the potting soil with straw, compost, leaf mold, or some similar material to prevent it from drying. Make sure the soil stays moist at all times when sprouting seeds.

To maintain the health of your plants, water every few days, or every other day if it is very hot. Watering is essential to keep an eye out for in your veggie container garden. Check your vegetables frequently to see if the potting soil has dried out. Check the soil by sticking your finger into it. If it seems dry below the surface level, water is needed. The best times to water are in the morning or evening, not in the heat of the afternoon. If desired, you can install a simple a drip irrigation system to make watering your container garden easier. It can water your vegetables for you automatically.

Keep an eye out for weeds and pests to ensure that your vegetable container garden remains at its most productive. While vegetables cultivated in containers are less susceptible to diseases than those planted in the ground, you should still keep an eye out for issues. Any plants that exhibit disease or major insect damage should be removed or treated. Since a container garden covers a small area, insect issues are usually able to be addressed by removing the insects by hand.

Generally speaking, you should replace the soil in your potted plants every two years. The soil's quality determines this. Every year, you should assess the potting soil's condition. You should replace the soil if you water the plants, and the water flows through the soil very quickly. You should replace the potting soil if you notice that it has grown hard to the touch and is not absorbing water properly. Additionally, you could see that the potted plants have become more prominent and are falling over. This signifies that the plants need to be repotted in new potting soil in a bigger pot.

Spring is the perfect season to replace the soil in the pots. Perennial plants are concentrating on developing new roots and foliage at this time. Since they can adapt well during the growth phase, altering the soil will not put them under as much stress during this time. Ideally you can replace the soil in the pots before you even plant your annual vegetables. For perennials this isn't an option, and the plant will need to be removed to freshen the soil.

To change the soil when there is an existing plant, remove the plant and root ball from the container. This is simple to perform if you have a plastic pot. Tap the pot on all sides until the plant comes out. Discard spent soil and partially fill the container with fresh potting soil until the plant can be placed in it. The plant's base should only be a few inches below the rim when the root ball is placed in the new pot. Fill in with fresh potting soil around the plant. Thoroughly water the plant in.

Planting Ideas for Container Gardens

The most enjoyable stage is of course, harvesting your home grown vegetables. Follow directions for the type of vegetable at hand, but in general, harvest vegetables as soon as your crops are large enough for you to appreciate them. If you harvest most vegetables frequently and early, they will produce more. Fruit set is often reduced when plants are not harvested early enough and are allowed to "go to seed."

It's a good idea to use pruners, scissors, or a knife to take what you need when harvesting anything other than root crops. If you try to pick off leaves or fruits, you risk hurting the plant and even uprooting it from the container.

The fundamental guidelines for producing a variety of vegetables in containers are provided here. Keep in mind that the recommended planting methods are for optimum growth. Vegetables may frequently be grown successfully in smaller containers.

- Direct seed beets directly into a 2- to 5-gallon window box.
- Transplant one cabbage plant per 5-gallon container. Alternatively, one plant per gallon container for a small variety.
- Direct seed into a 2- to 5-gallon container for carrots. Thin to about 3 inches apart.
- Direct seed or transplant two cucumber plants per 5-gallon container. Grow on a trellis or cage.
- Transplant one eggplant plant per five-gallon container.

- Directly sow green beans into a 5-gallon window box.
- Direct seed kohlrabi into a 5-gallon container. Thin to three plants.
- Direct seed or transplant lettuce into a gallon-sized or larger container. Thin to eight inches apart.
- Direct seed onions into a gallon or larger container. Green onions should be separated by two inches. Spread out bulb onions to a distance of five to six inches.
- Direct seed peas into a 5-gallon container. Use trellis for pole varieties. Thin to five inches apart.
- Transplant one pepper plant into a 5-gallon pot.
- Direct seed radishes into a 2-gallon or larger container. Thin to about three inches apart.
- Direct seed spinach into a gallon-sized or larger container. Thin to about three inches apart.
- Direct seed or transplant one summer squash plant per 5-gallon container.
- Direct seed or transplant four Swiss chard plants into a 5-gallon container.
- Transplant one tomato plant into a 5-gallon container. Grow on a trellis or cage.
- Direct seed one winter squash plant into a 5-gallon container. Grow on a trellis.

Home Compost

Compost is an organic substance that can be added to the soil to promote plant growth. Food scraps and yard waste account for more than 30 percent of what the average household discards and could be composted instead. Composting diverts these items from landfills, where they would use space and generate methane, a potent greenhouse gas.

All composting requires three fundamental components:

- Browns/Carbon - This category consists of natural materials such as fallen leaves, branches, and twigs. Plain cardboard (not coated and painted) and newspaper can also be used.
- Greens/Nitrogen - This category includes grass clippings, vegetable waste, fruit leftovers, eggshells and coffee grounds.
- Water

Your compost pile should contain an approximately equal amount of brown and green materials. Additionally, you should alternate layers of organic materials with particles of differing sizes. The brown materials offer carbon, the green materials provide nitrogen, and the water provides moisture to aid in the decomposition of organic matter.

Materials that can be composted:

- Fruits and vegetables
- Broken eggshells

- Grinds from coffee and filters
- Tea bags
- Nut husks
- Shredded newspaper (no painted inserts)
- Shredded cardboard (not coated or painted)
- Yard trimmings (not treated with chemicals)
- Grass clippings (not treated with chemicals)
- Houseplants
- Forage and straw
- Mulched leaves
- Sawdust
- Wood flakes
- Fur and hair
- Ashes from the fire
- Chicken, rabbit, or livestock manure

What Should Not Be Composted and Why:

- Black walnut leaves or branches
    - Emits chemicals that may be toxic to plants
- Coal or charcoal ash
    - May include plant-harming chemicals.
- Eggs and dairy products (butter, milk, sour cream, yogurt, etc.)
    - Cause odor problems and attract rodents and flies as pests
- Plants with diseases or pest infestations

- - Diseases or insects could potentially survive and be transmitted to other plants
- Fats, lard, grease, or oils
  - Cause odor problems and attract rodents and flies as pests
- Meat or fish leftovers and bones
  - Cause odor problems and attract rodents and flies as pests
- Pet waste (i.e., dog or cat excrement and soiled kitty litter)
  - May contain hazardous parasites, bacteria, germs, pathogens, and viruses
- Chemically treated grass and yard trimmings
  - May harm beneficial composting organisms

Advantages of Composting:

- Enriches soil with nutrients.
- Aids in water retention and plant disease and insect suppression.
- Reduces the requirement for synthetic fertilizers.
- Promotes the growth of beneficial bacteria and fungus that decompose organic debris to produce humus, a nutrient-rich substance.
- Reduces landfill methane emissions and reduces your carbon footprint.

There are numerous methods for constructing a compost pile. A water hose with a spray head, a pitchfork, square-point shovel, and machete are helpful tools. Regular mixing or turning of the compost, along with a small amount of water, will aid the compost in breaking down more quickly.

Choose a dry, shady location close to a water source for your compost container or pile. You can purchase or build an enclosed compost bin if desired, or you can literally just start a pile on the ground. It is ideal to locate your compost bin or pile in a flat, well-drained area.

As brown and green items are collected, add them to the pile, ensuring that more significant bits are chopped or shredded. Aim for a mix of about half brown and half green items. Wet dry substances as they are introduced. Optionally, you can cover the compost's surface with a tarp to keep it damp. When the material at the bottom becomes dark and nutrient-dense, the compost is ready for use. Typically, this takes between two months and two years, depending on the climate and how often the pile is mixed.

It is ideal to have an under the sink compost bin. You can purchase this at a local hardware store, gardening supply store, or create one yourself from a recycled container. This way, you don't have to walk outside to the compost pile every time you have something to add. You can just go out when the under the sink bin gets full.

The key to producing quality compost is a balanced mixture. You must maintain a healthy balance of 'greens' and 'browns' in your compost. If your compost is too moist, add additional "browns." If the mixture is too dry, add some "greens." Also essential is ensuring that the mixture contains sufficient air. Adding bits of crumpled cardboard is a simple technique to produce air pockets that will help maintain the health of your compost. Additionally, air can be introduced by mixing the components.

If you have issues with compost taking a long time to break down, ensure that the materials you use are chopped up and that you are mixing the pile regularly. Although it's not usually necessary, by using a compost activator, you may stimulate the proper enzymes in your compost. These products can expedite the transformation of grass, leaves, and garden debris into black, rich, crumbly compost. You combine a little quantity of the activator with water, then sprinkle it into your compost. After about 10 weeks of decomposition, the compost should be ready for use. Compost activator can also be used to revitalize partially decomposed or decomposing piles.

Fallen leaves are an excellent supply of compost. If you have a significant quantity of leaves, you may opt to place them in a sizeable biodegradable leaf bag rather than your compost container. After collecting fallen leaves, they can be left to decompose into an excellent source of moisture-rich soil amendment referred to as "leaf mold". The leaves will be contained in a single location, and the bag will biodegrade, leaving you with a magnificent compost pile.

When your compost is ready, the bottom of your bin or pile will have a dark brown, nearly black soil-like layer. It will have a spongy consistency and be nutrient-dense. Spreading finished compost on gardens improves soil quality by retaining water and reducing weed growth. Additionally, it decreases the demand for chemical fertilizers.

Purchased Compost

When you purchase compost, the products used to create it will differ. A variety of organic components are used to create compost. Purchased compost will commonly be made from one or more of the following materials:

- Yard waste
- Forestry byproducts including bark dust and sawdust
- Crop waste, such as rice hulls, processed plant material, and straw
- Animal waste from cows, chickens, and other types of livestock
- Food leftovers from restaurants, breweries, vineyards, and other establishments
- Waste from mushroom cultivation
- Worm-produced compost (vermicomposting)

The quality of compost varies. The kind of raw materials influences the quality. How the compost is made also influences the quality. Compost of the highest caliber should resemble black topsoil. It should have of a thin, flimsy construction. Fine compost with small particles is best. Ideally, it should go through a 3/8-inch screen. Nothing like pebbles, rubbish, or other waste should be present. Large compost particles should be removed for vegetable gardens and lawns. Chunky compost is suitable for landscaped areas.

Compost of the highest caliber smells earthy, like forest dirt. Avoid purchasing compost that smells bad or like ammonia or sulfur. These odors show that the composting process is not finished or that the balance of browns to green was not right when the compost was being made. Before buying bulk compost, inquire about the vendor's policy on pesticide and herbicide contamination. If you don't get a satisfactory answer, avoid the purchase.

A Note on Fertilizers

Agriculture has been performed without synthetic chemicals for thousands of years. Midway through the 20th century, artificial fertilizers were first developed. These chemical fertilizers were inexpensive, potent, and transportable in mass. The new artificial fertilizer technique was advantageous in the short term but had significant long-term adverse effects, including soil compaction, erosion, and reductions in total soil fertility, as well as health issues over the entry of harmful chemicals into food.

There is abundant evidence that compost and conservation tillage produces comparable or greater yields than chemical fertilizers. Utilization of compost promotes maximum reliance on locally or farm-renewable resources. This inexpensive "fertilizer" improves soil structure, texture, and aeration, increases the soil's water retention capacity and stimulates healthy root development.

Processed organic fertilizers are generally much better than standard chemical fertilizers for the environment and your garden because of the ingredients used to create them. Typically, synthetic fertilizers contain compounds that are not readily biodegradable. These chemicals leak into the ground and eventually enter the water system, where birds and other wildlife consume them. Groundwater contamination from chemical fertilizers is common in agricultural areas. In contrast, organic fertilizers lack these harmful compounds and, as a result, pose less threat to the environment.

However, overreliance on organic fertilizers can also cause problems, and can be expensive. With proper soil health being maintained, many vegetable gardeners are able to add an organic fertilizer only once at the beginning of the growing season and use nothing but compost for the remainder of the plant's lifespan. Some gardeners have built their soil up to the highest quality and no longer need fertilizer at all, other than their own homemade compost or high quality locally purchased compost. With proper attention paid to soil health, processed fertilizers are generally unnecessary.

Getting Your Soil Ready for Winter

Preparing the soil for winter is necessary for year-round success in the home garden, whether you produce ornamentals or edibles. Soil provides the basis for remarkable tree and plant growth. We want to prepare the soil in the winter with the spring growing season in mind.

Large soil areas should not be cleared of weeds until after the first frost. This should assist in reducing labor. The weeds will also protect the soil from erosion. You don't want to leave soil bare. After the first frost, your work area will have begun its transition to winter dormancy, and tender annuals will have withered. This is the time to clear the garden of weeds for the winter.

Once you have cleared your garden space of weeds, the soil should be covered. Covering the soil over the winter assists in erosion prevention and feeds the soil back with organic matter. You can top your beds with your chopped up dead annual plants and fallen leaves at no cost. Weed free natural mulches such as straw are another option. Alternatively, you could sow a cover crop instead of mulching with organic materials. (In this case, time clearing the weeds and sowing the cover crop according to planting directions for the seed at hand.)

Tender perennials with shallow roots, such as strawberries, will benefit from a mulch of straw, although hardier perennials like horseradish may need minimal aid. Mulch is advantageous to both berry plants and fruit trees. This simple procedure protects the root system from wintertime soil temperature fluctuations. You can also mix annuals and fallen leaves into the topsoil, replenishing the soil with organic matter. Weeds may be eliminated or added to the compost pile.

Before mulching, examine the soil at the base of your trees, berry plants, and other perennial garden plants once you have completed the cleanup of your garden. If the soil around your trees or plants has deteriorated, fill the void with more soil. If the soil or mulch surrounding a mature plant has grown hard or compacted, consider stirring a small amount of compost or worm castings into the soil around the plant. Avoid disturbing or damaging any roots during the operation. Finish by applying a new layer of mulch.

Clean the area around the base of the trunks of trees or stems of any plants buried by trash or soil washed into the incorrect location. Leaves are an excellent addition to compost, but they should not be used as mulch around live plants unless they are well shredded, since they can become heavy and trap moisture around plants, which can cause rot. (Although, un-shredded leaves can be used as a winter mulch in unplanted garden areas.)

If you have planted numerous cycles of annuals, such as certain vegetable plants, in the same spot, the soil may lose its nutrients and helpful bacteria. While spring is the ideal time to apply fertilizer, spreading compost after fall cleanup can revitalize depleted soil and boost your garden in the spring. At least two to three inches of aged compost should be applied to the soil's surface for the most significant benefit. Or, if you are using raised beds, just top them off.

Before the ground freezes, fall is the ideal time to prepare your soil for winter, so take advantage of the loose, workable composition and clean up your garden. In addition, autumn is an ideal time to have a soil's Ph tested to apply lime or sulfur if needed. Doing it now allows you time if you need to correct the soil pH before the next growing season.

Cover Crops

Since the time of the Roman Empire, planting cover crops has been a traditional and profitable agricultural technique. Since then, the strategy has been widely implemented in agriculture, resulting in numerous benefits. Farmers enjoy a profusion of cover crop benefits that suit a variety of aims, both in the short- and long-term. You can enjoy these same benefits in your vegetable garden.

Farmers plant cover crops to enhance soil quality for a variety of reasons. Cover crops contribute to healthy soil and fertility, reduce soil erosion, regulate soil moisture, attract pollinators, aid in weed and insect control, provide mulch and a source of green manure and organic matter, and are often utilized as cattle grazing or feed. Nitrogen is added or removed depending on the type of cover crop.

Cover crop roots help avoid wind and water erosion. Some species can convert nutrients into more digestible forms for other plants. Farmers obtain mulch as part of integrated weed management by planting cover crops between rows. Moreover, species that bloom between rows attract pollinators. Cover crops with shallow and deep roots penetrate the soil at varying depths, combating compaction and enhancing aeration.

Nitrogen enrichment rises with legumes and decreases with non-legumes. Legumes are regarded as cover crops that produce nitrogen. However, they do not capture and transform gaseous molecular nitrogen from the atmosphere on their own. Cover crops that fix nitrogen have a symbiotic relationship with the soil-dwelling rhizobia bacteria on their roots (formerly classified as agrobacterium). The bacteria, in turn, obtain carbohydrates from the legumes. Most nitrogen is released when nitrogen-fixing covers crops or microbes decompose. Rhizobium also reduces root problems in plants. The non-symbiotic cyanobacteria family provides a second biological option for converting air nitrogen into soil nitrogen.

Cover crops can absorb extra water after winter precipitation, improve water penetration (and soil aeration) via their roots, and preserve moisture for later income crops. The choice of crop to plant depends on its type and sequence. For instance, legumes are helpful for future cash crops because they provide nitrogen. Consequently, gardeners should not sow them when nitrogen levels are at their maximum. The time required for proper cover plant decomposition is an additional factor. For instance, buckwheat residue decomposes more quickly than barley or sorghum residue.

In agricultural practice, farmers sow cover crops at different times, either in the fall, late spring, or summer, in rows or between the crops. Some cover crops are naturally winter killed, while others must be removed, and their residues managed. Farmers may sow a single species or a mix of cover crop species.

Grasses, legumes, and broadleaf non-legumes are the three primary types of plants used as cover crops, based on their features and usage options. In most situations, they simultaneously serve multiple purposes, such as reducing erosion, enhancing soil quality, and providing grazing.

Grasses include annual grains such as buckwheat, rye, wheat, corn, barley, and oats. They grow relatively quickly and produce wastes that are readily controlled. Their fibrous, threadlike root systems are resistant to erosion and provide stability. Regarding nutrients, they collect soil nitrogen through a symbiotic relationship with a beneficial bacteria called Azospirillum, but they cannot fix atmospheric nitrogen.

As nitrogen-fixing cover crops, legumes are well-known for nitrogen enrichment. When plants grow large, their robust taproot system aids in combating undesirable subsurface compaction. Additionally, the larger the plant, the more nitrogen it can fix. Red and white clover, cowpeas, alfalfa, hairy vetch, and fava beans are examples of legumes.

Non-leguminous broadleaf plants absorb nitrogen from the soil, hold it in place, and produce green manure. They typically perish in the harsh winter weather without additional intervention. This category consists of brassicas, forage radishes, turnips, marigolds, and mustards.

When planting winter agricultural cover species, gardeners should consider their frost resilience. Winter hardy plants are resistant to extreme cold, whereas winter killed plants are sensitive to significant temperature drops and perish during winter. A winter hardy cover crop has the benefit of producing green manure in the spring but will be more work to incorporate than a cover crop which is naturally winter killed. If your cover crop is winter killed and your soil appears to have bare spots, be sure to cover it with organic materials such as fall leaves or straw, as discussed previously.

In raised beds and no dig gardens, winter hardy cover crops that have not been naturally winter killed can be cut at ground level and chopped up or can be mowed or cleared with a clearing sickle as low as possible, depending on the species. This should be done 6-8 weeks before planting the area and before the cover crop goes to seed. This process provides green manure which will be a rich source of nutrients for your crops and does not damage the soil's ecosystem as happens when cover crops are tilled in.

In the case of grass type cover crops, it may be necessary to cover the area well with overlapping newspaper or cardboard after mowing to prevent the grasses from growing back. Legume and non-leguminous broadleaf types of cover crops are more low maintenance than grasses. Grasses can be beneficial as a cover crop in gardens with compaction issues as the roots go deep and provide aeration, but this should not be a problem in raised bed and no dig gardens.

Chop and Drop

The chop and drop system is an easy, straightforward technique to improve your soil. The chop and drop approach involves chopping wasted crops, such as annuals towards the end of the season and letting the foliage fall to the ground. The plant remnants protect the soil by covering it and eventually decompose into compost at ground level.

Chop and drop mimics the natural process of plant decomposition. This makes the process of compost creation easy and natural, requiring less time and energy to develop healthy soil and benefit crops. Conventional compost systems (or purchased compost) are usually still necessary but chop and drop is perfect at the end of the growing season when you are preparing your soil for winter. The crops grown there are used to protect and feed back the soil where they grew.

Here are the simple steps to adopt chop and drop in your garden:

1. Cut annual plants down to their roots and prune perennial plants and trees.
2. Place the trimmings (large, chopped pieces) directly on the ground rather than transporting them to the compost bin.
3. The roots of the cut annual plants will remain underground, where they will stay rather than being dug up. The roots will help aerate the soil and eventually will rot and feed the soil.

4. Optionally, you can spread a thin layer of compost or manure over the plant materials. The best possibilities for manure are chicken, sheep, and cow manure, as they do not require as much time to age as horse manure. Rabbit manure can also be used and is the only type of manure that does not need to be aged.
5. Optionally, you can cover the plant material with a mulch layer. This will aid in the decomposition of the waste and plants. It will also prevent wind and precipitation from transporting them away. Pea straw is an excellent option for mulch, but any natural mulch besides wood chips will do. Mulch is an especially good idea if you did not have enough chop and drop materials to cover all of your soil, as you do not want to leave any soil bare.

The most evident advantages of chop and drop gardening are its simplicity of implementation and effectiveness in providing nutrients back to the soil. It also acts as a mulch that protects the soil. Additionally, chop and drop helps to aerate the soil. As a result of the plant roots decomposing beneath the soil, drainage is enhanced, and the soil is loosened. Thus, the soil is less compacted, increasing its fertility. The roots also transport organic matter deep into the soil, making it more accessible to future plants.

Chop and drop gardening attracts earthworms, which further aid in soil aeration. As earthworms consume the decaying plant materials, they burrow into the soil and deposit the nutrients there. This plant debris is transformed into worm castings, which are extraordinarily beneficial to soil health.

Chop and drop is one of the most natural methods of soil restoration. Nature has been growing plants successfully since the beginning of time, so mimicking the process is a simple way to achieve your gardening objectives.

Most plants are simple to cut and dispose of. Vegetables, annual flowers, and trees can all be incorporated into the chop and drop technique. The greater the plant material, the longer it will take to decompose. The largest materials such as very thick plant stems and larger tree branches should not be chopped and dropped because of the long time it would take for them to decompose.

The chop and drop process can be accelerated by chopping the plant tips. For instance, small tree branches are most valuable when they are chopped up. In many cases, gardening shears can chop plant material into smaller pieces. Additionally, placing the plant tops in a pile and mowing them is beneficial.

Dynamic accumulators, such as comfrey, are regarded as some of the most excellent chop and drop plants due to their capacity to take nutrients from the soil's depths and store them in their leaves. When these plants are chopped and discarded, the nutrients they contain become accessible to shallow-rooted plants in the topsoil.

You may want to chop and drop plants before they go to seed or take the time to remove and save the seeds. If you chop and drop vegetable plants that have gone to seed they may sprout. If you want volunteer plants, this is not a problem, but if you adopt crop rotation, these spring seedlings can be a nuisance.

If you chop and drop healthy plants, you shouldn't be concerned about transmitting diseases to subsequent crops. The plant matter will decompose like it would in a compost bin. Do not chop and drop plants that were diseased. The safest option is to dispose of these unhealthy plants in the trash or to burn them.

The fantastic thing about the chop and drop technique is the simplicity of it all, and you don't need much. However, a few gadgets do make the job easier. Primarily, you will need a quality pair of pruners. This is the tool you will use most frequently.

If you are chopping and dropping from fruit and nut trees, you may discover that something more robust than pruners is required. In this instance, it would be helpful to have loppers on hand. If you can find a pair of pruning shears with extending handles, you will be able to reach higher branches and have more leverage for easy cutting.

A clearing sickle is helpful if you have grown cover crops in a larger gardening area and you are chopping and dropping them for the soil's benefit. This gadget makes short work of lowering dense cover crops to the ground and prevents the need for mowing.

Chop and drop is a simple yet effective approach to enriching the soil. If you let nature take its course, you will have significantly less work to do.

Conclusion

The majority of gardeners jump straight into the task without taking the time to understand the base of their garden, the soil. They focus on the plants and what they require, but don't think about the soil and what it requires.

You have taken an important step to becoming a better gardener by learning about soil science. By understanding the soil's unique ecosystem and understanding how to maintain its structure, health, and fertility, you have set yourself up for abundant harvests in your vegetable garden from now into the future. Rather than being degraded over time, your soil will be regenerated over time and get better and better in the years to come.

Happy gardening!

YOUR REVIEWS HELP INDEPENDENT AUTHORS!

If you enjoyed this book and learned a lot about soil science, please leave a Five Star review on Amazon. Thank you!

James Bright

Printed in Great Britain
by Amazon